SpringerBriefs in Ecology

SpringerBriefs present concise summaries of cutting-edge research and practical applications across a wide spectrum of fields. Featuring compact volumes of 50 to 125 pages, the series covers a range of content from professional to academic.

Typical topics might include:

- A timely report of state-of-the art analytical techniques
- A bridge between new research results, as published in journal articles, and a contextual literature review
- A snapshot of a hot or emerging topic
- An in-depth case study or clinical example
- A presentation of core concepts that students must understand in order to make independent contributions

More information about this series at http://www.springer.com/series/10157

Ian D. Rotherham

Recombinant Ecology - A Hybrid Future?

 Springer

Ian D. Rotherham
Faculty of Development and Society
Sheffield Hallam University
Sheffield
UK

ISSN 2192-4759   ISSN 2192-4767   (electronic)
SpringerBriefs in Ecology
ISBN 978-3-319-49796-9   ISBN 978-3-319-49797-6   (eBook)
DOI 10.1007/978-3-319-49797-6

Library of Congress Control Number: 2016958471

This Springer imprint is published by Springer Nature
The registered company is Springer International Publishing AG
The registered company address is: Gewerbestrasse 11, 6330 Cham, Switzerland

Photograph: Canada goose brought in to adorn great landscape parks of the 1700s and 1800s

Foreword

When I first arrived in Australia in 1970, I was excited to explore vegetation, flora and fauna I had read about but had never seen 'up close and personal'. In addition, I was determined to spend time researching the unique qualities of the diverse and complex Australian ecosystems. Yet after a couple of years, I realised the ecological challenges (and perhaps opportunities) lay in ecosystems which appeared as rough and ready weedy places, with invasive plants from all continents mixing it with native species. Sometimes, these invasive mixtures simply overpowered native species, at other times there was obviously a clear balance between native and invasive, alien, adventive—all negative terms ascribed to species that simply 'do not belong'. In the bushland of the rapidly expanding City of Perth, I found not only some of the richest flora I had ever encountered, but also non-native species apparently thriving within an existing ecosystem.

One such plant is *Gladiolus caryophyllaceous*—an endangered species in its native South Africa, but regarded by many as a potential threat in Western Australia. Whilst it can clearly become quite invasive in some circumstances, at low levels of colonisation it is simply an addition to the native communities. There are other species rather more problematic for Western Australian bushland, for example *Erhharta calycina,* the perennial veldt grass also from South Africa, can invade and destabilise native vegetation. To try and understand why these new ecologies were happening I visited South Africa—and there discovered the vigorous invaders in Australia were much less vigorous components of their native ecosystems—in contrast with many Australian species which were creating vegetation types previously unknown, or in some cases actual replicas of systems in similar climates in Australia.

From that point, I went on to observe this phenomenon in many places and coined the term 'synthetic vegetation' for such new ecologies. This term was designed to reflect the synthesis evident in, e.g., the new forests growing up in disused railway sidings in Britain. In these forests, a canopy of *Betula* and *Buddleia* develops over a rich grass and herb understorey of both native and non-native species. That term, however, did not resonate well. At about the same time, Michael Soulé in the USA had coined the term recombinant ecology, which has also had mixed uptake, but in this volume, it is successfully employed by Ian Rotherham. The publication by Hobbs et al. (2006) on 'Novel ecosystems: theoretical and management aspects of the new ecological world order', marked a step change in focus on the realities facing ecosystems globally—no less in Britain. Whilst there are still many who doubt the appropriateness of publishing and talking on the topics of novel ecosystems, recombinant ecosystems, synthetic vegetation, or other titles, given to this group of emerging ecosystems, that is a myopic view of where we are today. Moreover, where we are is in the Anthropocene, and that epoch is yielding many challenges for us as a species, but also huge opportunities.

In the pages that follow, Ian Rotherham sets out an agenda for understanding and managing recombinant ecologies. That there is much to think about in recombinant ecology is explained these writings, not least how we live with and manage these new systems. For, despite the noise and clamour around rewilding, they will be our new wild places, they will offer homes for species endangered in their original locales, and they will create conditions, which will allow many species to flourish in old and new combinations. And, of course, some species will be threatened by this new ecology—but they may well have been under threat from environmental change already.

Finally, whilst this volume has an especially British flavour, accelerating globalisation of biodiversity means its conclusions and observations will be helpful everywhere, in contributing to the debate on the new ecological world order.

June 2016   Peter Bridgewater
                                                                             Adjunct Professor
                                                                Institute of Applied Ecology
                                                                   University of Canberra
                                                                                   Australia

Reference

Hobbs RJ et al (2006) Novel ecosystems: theoretical and management aspects of the new ecological world order. Glob Ecol Biogeogr 18:1–7

Photograph: Fallow deer—a long-term introduction by the Romans and then the Normans, now thoroughly established and often a keystone species

Preface

As ecologists, for over a hundred years, we have sought bold, overarching ideas and theories to provide insight into big issues and the 'bigger picture' of ecological systems and processes. From Tansley's seminal volumes such as 'The British Islands and their Vegetation' (1949), to Rodwell (ed.) (1991a, b, 1992, 1995, 2000) with the National Vegetation Classification, ecologists have attempted to frame the national ecology systems into a logical, rational system. There is a tendency, however, for such approaches to try to capture some form of 'correct' stasis of condition, and this, perhaps, is an inherent weakness. Others, such as Grime et al. (2007), Grime and Pierce (2012), or Allen and Hoekstra (1992), aim to understand the 'bigger picture' of processes, dynamics, and the biological nature at the core of ecological systems.

This book addresses critical issues of the changing nature of ecology and ecosystems consequent on urbanisation, globalisation, climate change, and human cultural influences. From long-term human interactions through nature in agriculture and forestry, to increasingly major impacts of urbanisation and other environmental changes, people have forced and facilitated the hybridisation of nature. Indeed, in the face of human-induced and natural climate changes as globalisation accelerates, the pace of this hybridisation speeds up. Anthropogenic influences cause disturbance, nutrient enrichment, habitat replacement (through formation and destruction), and dispersal of species on a planetary scale. The ecological processes that drive the changes are the 'natural' mechanisms of ecological successions and changes, and of species and ecosystem hybridisation or adaptation. Today though, the mixing of species is occurring at a rate that is unprecedented in the history of biodiversity evolution. The so-called Anthropocene, the latest great evolutionary epoch is upon us and nature is adapting to a new canvas and a changed template. Issues relating to this recognition are discussed in a popular volume by Davies (2016) and seem to be directly relevant to recombinant and future ecologies.

The dramatic and largely unrecognised consequence of these human influences is a hybridisation of both species and of ecology itself. Whilst this process is most easily observed and recognised in the increasing urban environments of the planet, it occurs more widely, such as in forestry and in agricultural landscapes. As new environmental conditions are forged, plants, animals, and fungi move and mix, beyond their natural distributions and limits, old and new, native and exotic, become enmeshed in recombinant communities and hybrid ecosystems. Here, and especially in the rapidly expanding urban heartlands of this new ecology, native and alien jostle for position with novel interactions and dependencies are formed.

This short volume brings together key research for the first time and considers the implications for future conservation. The issues of alien invasive species and natives are controversial and raise serious conservation and economic concerns (McNeeley 2001; Simberloff 2011; Smout 2011). Many human aspirations for nature conservation and our subjective perceptions of what is 'good' are based on the ideas of stability and continuity. However, it is increasingly clear that whether or not we like it, the world is not stable and indeed, is getting less so. Continuity is vitally important for nature conservation and land-associated management, but stability is problematic. As I discuss elsewhere (see Rotherham 2014, 'Eco-History'), a serious problem is not necessarily that of stability per se, but of the nature and scale of human ecological impacts and disturbance, as we move from micro-disturbance in pre-industrial societies to massive macro-disturbance and eutrophication with large-scale industry and urbanisation. The impacts of the changes are profound, and this important publication is the first step on the way to a fuller recognition of what they imply for a future ecology. Whether we like or not, our emerging ecology will be hybrid, and indeed, history suggests that it always has been. Some years ago, Thompson et al. (1995) stated that we can expect a continuation of the spread and invasion by species around the world, but that attempts to understand such phenomena should be placed within the existing frameworks of species range dynamics. Particularly, as they noted, invasions and the characteristics of invaders are '... of crucial importance to global change and declining biodiversity'. Furthermore, they noted how research at the time had attempted to identify the key ecological characteristics of invasive species and potentially invisible ecosystems, citing the example in the literature of Lodge (1993). Other attempts to address these issues include, for example, Noble (1989). However, they questioned the assumption that '... there is something unique about successful invaders, in the sense of species which establish themselves beyond their native range'. Thompson and colleagues suggested that '... there is nothing unnatural or even unusual about changes in species' ranges'. The assertion is supported by the observation that '... the entire biota of north temperate regions has migrated large distances within the last 10,000 years, and detailed recording reveals a constant ebb and flow of species' ranges'. They note how invasions are '... distributed along a continuum from natural expansions in distribution at one extreme to human-mediated transfer between continents at the other'. An important point too is that 'invasions' are generally perceived as movements across national borders, even though these are frequently of little biological significance. This issue of borders resonates with the thoughts of historian, Charles Warren (2005), when considering the impacts on attitudes to alien invaders of devolution in countries such as Scotland.

A context for these issues is a large and rapidly expanding literature on alien and invasive species around the world and focussing mostly on the undoubted issues and challenges that they raise (e.g. Simberloff 2003, 2005, 2011; McNeeley 2001, 2011). This is reflected in the publications focussed on the British flora (e.g. Stace and Crawley 2015). Invasive pests and diseases also have an extensive body of published research and popular reviews (e.g. Alford 2011); this is a frequently unappreciated aspect of invasions and new ecological processes and consequences. There is also a body of work on distinctively urban ecologies though much of this is on the natural history aspects (e.g. Goode 2014) and rather less on the underpinning ecological aspects such as the synurbic character of distinctly 'urban' species (e.g. Francis and Chadwick 2012). However, there seems to be a significant gulf between the works emerging from say Australia, New Zealand, and North America such as Hobbs et al. (2013), emphasising the development of 'novel ecosystems' on 'hybridisation', 'recombinant ecologies' (see Meurk 2011), and the British literature. Indeed, in terms of addressing the matters of recombinants, the latter has barely moved on from the work of Barker (Barker 2000; Barker et al. 1994). On the other hand, there is a significant output on more popular interest in the cultural histories of exotic species (e.g. Mabey 1996), and in popular scientific writing on new ecologies (e.g. Pearce 2015). Much of the work on novel ecologies relates strongly to concepts of supposedly pristine, native ecologies that have been degraded and to the extensive literatures on restoration ecology, for example. As such, the ideas and concepts help inform our view of the European and British situation, but the application and the language do not always transfer easily to landscapes dominated by long-term cultural ecologies and those highly modified by intensive land use or urban sprawl.

Here, I deliberately focus on the example of the British Isles, which benefits from a uniquely long timeline of ecological research and a good understanding of relevant aspects of environmental and social history. However, there is also the benefit of considering the intimate ecology of a group of islands, and indeed, a biogeographic area that has received species imports and colonisers from overseas for many centuries. I hope that this choice of case study will provide much of interest to other regions too.

In Chap. 5, I touch on 'climate change and ecological hybridisation', but do not dwell in detail on this theme. This is a short book, and that is a big topic for another occasion.

Sheffield, UK   Ian D. Rotherham

References

Alford DV (2011) Plant pests. Collins new naturalist, London
Allen TFH, Hoekstra TW (1992) Toward a unified ecology, 2nd edn. Columbia University Press, New York

Barker G (ed) (2000) Ecological recombination in urban areas: implications for nature conservation. English Nature, Peterborough, pp 21–24. Barker G (ed) (2000) Ecological recombination in urban areas: implications for nature conservation: a workshop held at the centre for ecology and hydrology (Monks Wood), 13 July 2000. UK Man and Biosphere Committee, Urban Forum, English Nature, Centre for Ecology and Hydrology, Peterborough, ISBN 185716542X, 9781857165425, 25 p

Barker G, Luniak M, Trojan P, Zimny H (eds) (1994) Proceedings of the second European meeting of the international network for urban ecology. In: Memorabilia Zoologica, vol 49. Warsaw

Davies J (2016) The birth of the Anthropocene, University of California Press, Oakland, California

Francis RA, Chadwick MA (2012) What makes a species synurbic? What makes a species synurbic? Appl Geogr 32(2):514–521

Goode D (2014) Nature in towns and cities. Collins New Naturalist, London

Grime JP, Pierce S (2012) The evolutionary strategies that shape ecosystems. Wiley-Blackwell, Chichester

Grime JP, Hodgson JG, Hunt R (2007) Comparative plant ecology. A Functional approach to common British species, 2nd edn. Castlepoint Press, Dalbeattie

Hobbs RJ, Higgs ES, Hall CM (eds) (2013) Novel ecosystems. Intervening in the new ecological world order. Wiley-Blackwell, Chichester

Lodge D (1993) Biological invasions: lessons for ecology. Trends Ecol Evol 8:133–137

Mabey R (1996) Flora Britannica. Sinclair-Stevenson, London

McNeeley JA (2011) Xenophobia or conservation; some human dimensions of invasive alien species. In: Rotherham ID, Lambert R (eds) Invasive and introduced plants and animals. Human perceptions, attitudes and approaches to management. Earthscan, London, pp 19–38

McNeeley JA (ed) (2001) The great reshuffling. Human dimensions of invasive alien species. IUCN, Gland, Switzerland

Meurk C (2011) Recombinant ecology of urban areas: characterisation, context, and creativity. In: Douglas I, Goode D, Houck MC, Wang R (eds) (2011) The Routledge handbook of urban ecology. Routledge, London, pp 198–220

Noble IR (1989) Attributes of invaders and the invading process; terrestrial and vascular plants. In: Drake JA, Mooney HA, di Castri F, et al. (1989) Biological invasions a global perspective. Wiley Chichester, pp 301–313

Pearce F (2015) The new wild. Why invasive species will be nature's salivation. Icon Books, London

Rodwell JS (ed) (1991a) British plant communities: volume 1, woodlands and scrub. Cambridge University Press, Cambridge

Rodwell JS (ed) (1991b) British plant communities: volume 2, mires and heath. Cambridge University Press, Cambridge

Rodwell JS (ed) (1992) British plant communities: volume 3, grassland and montane communities. Cambridge University Press, Cambridge

Rodwell JS (ed) (1995) British plant communities: volume 4, aquatic communities, swamps and tall-herb fens. Cambridge University Press, Cambridge

Rodwell JS (ed) (2000) British plant communities: volume 5, maritime communities and vegetation of open habitats: maritime communities and vegetation of open habitats. Cambridge University Press, Cambridge

Rotherham ID (2014) Eco-history: an introduction to biodiversity and conservation. The White Horse Press, Cambridge

Simberloff D (2003) Confronting introduced species: a form of xenophobia? Biol Invasions 5: 179–192

Simberloff D (2005) Non-native species do threaten the natural environment. J Agric Environ Ethics 18: 595–607

Simberloff D (2011) The rise of modern invasion biology and american attitudes towards introduced species. In: Rotherham ID, Lambert R (eds) Invasive and introduced plants and animals. Human perceptions, attitudes and approaches to management. Earthscan, London, pp 121–136

Smout TC (2011) How the concept of alien species emerged and developed in 20th-century Britain. In: Rotherham ID, Lambert R (eds) Invasive and introduced plants and animals. Human perceptions, attitudes and approaches to management. Earthscan, London, pp 55–66

Stace CA, Crawley MJ (2015) Alien plants. HarperCollins, London

Tansley AG (1949) The British islands and their vegetation. Cambridge University Press, Cambridge

Thompson K, Hodgson JG, Rich TCG (1995) Native and alien invasive plants: more of the same? Ecography 18:390–402

Warren C (2005) The concept of alien and native species: time for a rethink? ECOS 26(3/4):10–18

Acknowledgments

The publishers and their editors are thanked for their forbearance over timescales and more. I wish to thank all those over the years who have helped stimulate and inform my thoughts and ideas on the nature of ecology and especially on the 'native/alien' paradigm. In particular, the late Oliver Gilbert working in urban Sheffield opened my eyes to some of the often unspoken issues. George Barker was another pioneer of distinctive urban communities, and Peter Shepherd has led the way on urban vegetation in Europe. Rob Lambert co-edited a major volume on alien and invasive species in 2011 and brought to the project a wealth of knowledge of environmental history. Chris Smout has always helped by challenging current thinking and highlighting critical issues (e.g. see Smout 2000), and others such as Peter Shaw and Paul Ardron have raised my awareness of the unseen exotics all around us. The work of Philip Grime and colleagues has proved useful and pragmatic when considering issues of future trends and successional directions. Finally, Richard Mabey's popular writing has frequently taken ideas where academics have feared to tread. Peter Bridgewater, Philip Grime, and Rob Francis are all thanked for generously 'volunteering' to read and comment on a draft manuscript. I am also grateful to Peter for the Foreword.

Reference

Smout TC (2000) Nature contested: environmental history in Scotland and Northern England since 1600. Edinburgh University Press, Edinburgh

Contents

Chapter 1
An Introduction to the Concept of Recombinant Ecology

Oh Brave New World

Miranda: O wonder!
 How many goodly creatures are there here!
 How beauteous mankind is! O brave new world
 That has such people in't!
Prospero: 'Tis new to thee.

William Shakespeare, *The Tempest*, Act 5, Scene 1, 181–184

Global Ecological Crisis

According to Eisenhauer et al. (2011), 'Mankind faces multiple anthropogenic global environmental changes, which are now large enough to exceed the bounds of natural variability. One of the most significant consequences of contemporary global change is the rapid decline of biodiversity in many ecosystems. This unprecedented biodiversity loss has generated concern over the consequences for ecosystem functioning and services, and prompted a multitude of studies.' For the discussion that follows, this is the stark context of the wider environmental landscape of the twenty-first century.

Indeed, whichever way you look at the Earth's planetary ecology in the twenty-first century, it is clear that we have serious problems to face; and furthermore, many of the issues are of our own making (see for example, Rotherham 2014a). However, whilst some or even many of the ecological crises seem insurmountable and indeed growing, there are other, more positive aspects. Sometimes we fail to see these and particularly we find it difficult to adapt to change. An altered ecology is troublesome for nature conservation, and climate change with associated extreme weather events is uppermost in the minds of politicians, the media and

© The Author(s) 2017
I.D. Rotherham, *Recombinant Ecology - A Hybrid Future?*,
SpringerBriefs in Ecology, DOI 10.1007/978-3-319-49797-6_1

many of the public. Undoubtedly, climate is changing and it seems that weather is more extreme. Clearly as well, human activities both make this worse and reduce the ecosystem's ability to mitigate and to adapt. Yet it is also obvious that climate has always changed and extreme events have come and gone over the centuries, but we remain surprised by storms, floods, droughts and the like. Then, whilst we experience changed climate through extreme weather events, humans persist in polluting, draining, degrading, urbanising and generally despoiling the environment. We are still surprised when things go wrong!

In terms of the actual ecology of our planetary system, we have, over thousands of years, cleared, drained, ploughed, grazed and altered almost all terrestrial systems. Similarly, through pollution and over-fishing, and the destruction or modification of vital coastal zone areas, we have damaged the seas. Freshwater environments have fared even worse! As a rule, in an increasingly eutrophic and disturbed world, competitors and ruderals increasingly displace stress tolerant and ruderal stress tolerant species, (see Grime et al. 2007 for example). Opportunistic species and those with catholic requirements spread and with the twin drivers of globalisation (Hulme 2006) and urbanisation, plant and animal communities change with extinctions, invasions, displacements, and successions. Into this melting pot of biodiversity, humans add species from around the world mixed both accidentally and deliberately, many failing to establish, some persisting and relatively few thriving. However, as we change the baseline environmental conditions, often the balance moves away from established native species towards opportunistic invaders, which are frequently but not always exotics. Predictable processes of ecosystem dynamics drive successional changes in space and time, with communities of plants and animals fluxing as species ebb and flow in distribution and abundance. The speed and extent of such ecological mixing has increased dramatically in recent decades, with urbanisation and globalisation. However, it is important to realise that this process is not new, but has been going on for millennia. Furthermore, and a conceptual problem for many ecologists, in order to understand the process and the outcomes, we must recognise that humanity is an integral part of nature and not merely an inconvenient intrusion. Agnoletti and Rotherham (2015) described the bio-cultural nature of landscape, and Rotherham (2015a), the eco-cultural interactions of people and ecologies. To assume that humanity is somehow apart from nature rather than immersed in it is to make a huge mistake.

In this context, it is surprising that more attention has not been afforded to this human-nature interface and to the ecological systems, which it produces, or influences. There is a considerable literature on pertinent aspects of the subject, but little that brings it together. Furthermore, as much academic research goes on in silos with a passing nod to multi-disciplinary and cross-disciplinary studies, the necessary cross-over between scientific ecology and social sciences like history, has happened a little late in the day. To understand better our possible future ecologies, it is best to understand where we are today, and especially, how we got here. In order to address the paradigms of recombinant ecology and of novel ecosystems, we need to understand invasion biology and the ecologies of invasive and exotic species. However, given the dramatic influences of urbanisation and the emergence of urban recombinant ecologies over recent centuries, we must add knowledge of urban environments and processes too. Furthermore, if we are to tease apart the nature of these 'new' ecological systems, then it is necessary to relate them to basic building blocks of ecological systems through species strategies and successional processes. Finally, and frequently missing in scientific ecological accounts is the influence of human history on these ecologies, and indeed, often, of ecology on human history.

Gillson (2015) has raised a similar issue in relation to the role of paleoecology to inform debates on environmental change and future ecology. She notes how the complex and dynamic nature of ecosystems leaves conservationists the problem of shifting targets in terms of their objectives. 'As conservationists grapple with issues of biodiversity conservation and sustainability in the face of climate change, habitat loss, extinctions, pollution and socio-economic transformation, there is a need for a long-term perspective that guides ecosystem managements and conservation in our human-dominated epoch, the Anthropocene.' Indeed, and noted by Gillson, is the fact that conservation seeks to preserve systems which are in constant flux and in practice represent moving targets. These issues were previously discussed by Pickett and White (1985), du Toit et al. (2003), Lindenmayer et al. (2008), and Gillson raises the important point that a longer-term understanding is required. As conservation moves, in the face of modern challenges, to the maintenance of resilience, of ecosystem processes, and vital functions, rather than merely selected population sizes for particular species and for stability in ecosystem conditions, the paradigm has shifted. Gillson and Marchant (2014) state that there is a need to understand the limits of ecological resilience and the tipping points at which fundamental and dramatic ecosystem reorganisation may occur.

A consequence of our history and our cultures is that whilst we view and judge the ecologies around us through the lens of objective science, it is not impartial. Scientists do not like to admit it, but we are human and make value judgments. Certainly, in nature conservation and particularly in issues of the control of invasive, alien species, impartiality and sometimes both science and history are jettisoned. A reason for this is of course the calamitous impacts of many invasive plants, animals, fungi and diseases around the world (see for example, McNeely 2001, and Simberloff in Rotherham and Lambert 2011). Millions of people have died due to invasions, economies have collapsed and ecosystems have been overwhelmed. All this is incontrovertible. Yet, as noted in contributions to Rotherham and Lambert (2011), judgements of species being native or alien, invasive or not invasive, a problem or not a problem, have frequently been misplaced. Additionally, aspirations of control or even eradication, of invasive alien species have frequently lacked substance, resources, and the necessary vision or response. In many cases throughout history, the changes happen and nothing will stop them. Furthermore, and often not acknowledged, is that in many situations, what begins as a dynamic and aggressive invasion is absorbed into a modified 'recombinant' ecosystem. Many invaders become less invasive over time.

As noted earlier, there is a vast literature on conservation and ecology, and increasingly on alien species, but little on the core focus of this short book. Space does not allow a full review of the relevant works on the diverse topics, which serve to inform this discussion, so in this case the selection has to be light touch rather than comprehensive. A brief diversion into recent popular conservation literature (e.g. Adams 2003 on '*Future Nature*'; Goldsmith and Hildyard 1986; Weightman 1990) suggests an absence of anything of substance on recombinants, on ecosystem hybridisation, or ecological fusion. Pearce (2015) provides a glaring exception to this general rule. Furthermore, until very recently (e.g. Hobbs et al. 2013a), there was almost nothing in the academic literature on 'novel' ecosystems. Some ideas began to emerge with the writing of Hobbs and colleagues (e.g. Hobbs et al. 2013b), and especially of Bridgewater (e.g. Bridgewater 1990).

Most popular texts deal with core conservation 'issues' and politics, and academic texts and papers cover the constituent science, but few if any, address exotic species as anything other than pests to be controlled and eradicated, or quirks of passing natural history interest. Yet this is in a situation where humans have changed the face of the Earth. As noted (by for example Rotherham 2014a), human impacts on global ecologies have occurred over millennia, but in recent centuries these have triggered what many consider to be a new and distinctive geological epoch—the Anthropocene. Human land-use, exploitation and consumption of resources, industrialisation, intensive agriculture, urbanisation, and more have transformed climate, biodiversity abundance and distribution, land cover, soils, and water. With cultural severance, the disconnection of local people from their environmental resources, the changes noted above have triggered the sixth great extinction combined with accelerating climate change.

Understandably, much early conservation from the nineteenth century onwards has focused on the preservation of landscapes perceived in some way as 'wild'. Indeed, this idea has gained a new fashionability today. However, 'wild' implied or implies a lack of people and of human impacts, with humanity, ecology separated, and distinct. Yet throughout history, people and nature have interacted and each is embedded in the other. Visions of future nature, informed by long-term environmental histories, must acknowledge our role in nature. Furthermore, in accepting our role in nature we must take responsibility for our actions and impacts rather than humanity being apart, separate and independent.

In stirring the pot of 'ecological soup', humanity has changed ecologies and trigged novel associations and permutations. Today more than ever, we live in a world of recombinant and hybrid ecologies and this will increase in the future. Recognising this does not diminish the importance of nature conservation though it raises issues and questions that need to be addressed. These processes may be seen as 'game changers' in terms of the scope, scale and challenges for future conservation, and to blithely ignore them is naïve and will compromise effective conservation actions.

When the spectre of global climate change began to be taken seriously, it generated visions of apocalyptic losses in biodiversity and mass extinctions (see Thomas et al. 2004 for example). These calculations were based on assessments and models of ecological niche spaces and their climatic envelopes. However, some of the predications have since been modified and moderated. In some cases, (see Thomas 2011, 2013), they are counterbalanced with the idea that the Anthropocene may generate new species to compensate for the losses. Such ideas are exciting indeed, but should be viewed with a degree of caution. The argument is that the previous five mass extinctions and episodes of massive global environmental change triggered sudden extinctions followed by rapid evolution of new species. The time-scales and processes in these great evolutionary times were radically different from the present, and the outcomes for the dominant organisms of the time (such as for example the dinosaurs), and hence for humanity today, were not good! In terms of realistic and meaningful contributions to nature conservation and to 'biodiversity', (the two often viewed as synonymous, but they are not), such newly evolved species will be trivial in the context of the human experience of ecology. However, we will examine some of the new ecologies and species later.

As noted by Maraun (2012) in reviewing Grime and Pierce (2012), the search for general patterns is still the most important challenge for ecologists. In commenting on Grime's well-known CRS strategy theory, he also states that making ecology a predictive science seems even more difficult. Indeed, this is why Grime's work to develop a more predictive tool of the universal adaptive strategy theory or UAST is a helpful backdrop to my discussion. CRS has been and is a very useful approach for many fieldworkers and theoreticians and can assist in informing discussions on recombinant and hybrid ecologies. Not all ecologists feel comfortable with the CRS triangle, but regarded as a 'tool' and not a 'truth' it is exceptionally pragmatic and useful. Additionally, the system evolved firstly in relation to plants and a major thrust of the recombinant ecological process is through the dispersal, mixing and spread of plant species. Fauna and fungi are also significant but vegetation is at the core of many recombinant systems. Maraun accepts that there is still a long way to go in developing the predictive tools needed to understand ecological change, and for recombinant ecology in particular, this is certainly the case. However, a first and vital step, is recognising the extent of recombinant and hybrid ecologies that already exist, and then accepting the inevitable trends towards ever more systems in the future. Understanding the 'bigger picture' and developing predictive tools to guide ideas of ecological change becomes not only more difficult, but also increasingly important.

Something missing in many of current arguments and debates on climate change and species is the scale of human alteration of land cover and condition. For ecology, maybe the problem is not so much climate change but the fact that we have fragmented and modified the land surface to a degree whereby species cannot move in order to stay within their ecological envelope within the time available. To resolve this, we need to heal the wounds of humanity's impacts and to work as a part of nature on the inside not as a spectator looking in. Accepting that much of our ecology is recombinant and hybrid, and that many ecosystems are significantly altered by humanity, is a good starting point. In attempting to do this we also have to accept major differences in approaches and terminologies between 'Old World' and 'New World' ecologists, and today, the divergence of research and writing about 'novel' ecologies between the two. An irony of course is that so-called 'New World' ecologies are in many ways more 'ancient' than those of the so-called 'Old World', and the interest in novel ecology and recombination in the former has come about largely due to the (mostly) recent collisions between the ecologies of 'Old' and 'New' following Western European imperialism and globalisation.

Setting the Scene—Ecological History, Aliens, Exotics and Invasives

A starting point for an understanding of the nature of British ecology must be the nature of its history and the changes over time. In this respect, there is a wealth of detailed science with for example for vegetation, a rich and long-standing literature (e.g. Godwin 1975; Walker and West 1970; Ingrouille 1995). There is similar knowledge of the history of many faunal groups too, and much of this demonstrates the changing and evolving nature of Britain's ecology (e.g. Harris and Yalden 2008).

An essential component of recombinant ecological communities is the 'alien' species, but this is a term frequently used but rarely defined. Therefore, it is important to consider what exactly this means, and this itself varies considerably. In Britain, being a complex of large and small islands, those species that are native, or not, and when and how they arrived here, has long been an obsession for ecologists, particularly botanists. Dictionary definitions suggest:

'Belonging to another person, place or family especially to a foreign nation or allegiance. Foreign in nature, character or origin' (1673)

'A stranger or a foreigner. A resident foreign in origin and not naturalised' (1330)

'One excluded from citizenship, privileges etc'. (1549)

'A plant originally introduced from other countries' (1847). (Friedrichsen 1990)

Clement and Foster (1994) used 'alien' in a broad sense to denote all novel plants whether or not they were believed to have arrived because of human activities. They include plants referred to by other authors as adventives, casuals, ephemerals, exotics, introductions and volunteers. Ellis (1993) wrote a very useful and pocket-sized introduction to invasive plants in Britain. He suggests that for many, an alien plant is essentially one that is not native. In this case, a native plant is one that arrived in Britain prior to the closure of the English Channel around 7–8000 years ago, and so an alien is a species arriving after such a date.

For plants in particular, the mode of arrival or of introduction is central to both definition and to an understanding of the issues. Ellis (1993) took alien plants to be those introduced by people, both deliberately or accidentally. Nevertheless, he also noted that in reality, and with a longer time perspective, most native flora could be considered alien invaders. This is due to the dynamic and fluctuating nature of vegetation in a landscape with long-term changes of key factors such as climate.

This latter point may become increasingly important in the years to come, a significant observation and one to hold for later in this discussion. For the conservationist it complicates the underlying priorities for contemporary conservation management. However, this recognition does emphasise that much important conservation management is based substantially on subjective human needs, opinions and priorities, not necessarily founded on ecological science.

The origins of alien species may be significant and for plants, they arise from a diversity of sources.

Cultivation:

- Agricultural seed
- Grass seed
- Garden escape
- Greenhouse escape
- Relic from cultivation
- Introduced.

Foodstuffs:

- Grain
- Bird-seed
- Animal feed
- Oil-seed
- Food refuse
- Spice.

Other commodities:

- Wool
- Cotton
- Coir
- Ballast
- Timber and wood products.

They can be 'casual', 'persistent', or 'established', and are often described as either 'introduced' or 'naturalised'. The latter implies a self-sustaining and expanding population. The status of long-term introductions to the British Isles has long been an issue for botanist and now conservationists. Many species arrived here a long time ago and botanists quite like them! To get around this problem of our subjectivity in response to species invasions, in the more recent British plant atlas, 'The New Atlas of the British and Irish Flora' (Preston et al. 2002), they were described as 'archaeophytes', which makes them in effect 'honorary natives'.

This interest in exotic species has been essential in the emerging concepts of recombinance. In this respect, invasive and exotic species have long been of interest to ecologists and natural historians (for example Grindon 1859, 1864; Dunn 1905; Salisbury 1961; Fitter 1945; Lever 1977; Ratcliffe 1984; and most recently, Stace and Crawley 2015). However, for many botanists in the past, non-native plants were generally rarities, of considerable interest and excitement, or more commonly establishing plants of little note whatsoever. Certainly, it was relatively rare for them to be considered a problem. The emergence of conservation worries surrounding naturalising exotic species is a relatively recent phenomenon, though the more aggressive invaders like Himalayan balsam concerned some Victorian botanists such as Joseph Hooker and the Manchester botanist Grindon (Rotherham 2005a).

Sir J.D. Hooker, aged 82 stated that Himalayan balsam was 'A terror to botanists, deceitful above all plants, and desperately wicked'. In the Manchester Flora (1859), Grindon described 'The Impatiens coccinea, a tall and weedy plant, with flowers of a dull red colour, is rapidly disseminating itself, growing like its congeners, where-ever a seed is dropped.' Then in his 'British Garden Botany' (1864 cited in Rotherham 2005b), he noted 'The *Impatiens glanduligera*, a tall and weedy plant, is common in gardens, and fast disseminating itself over the country.' This gradual recognition of the invasive nature of such introduced plants is discussed further in Rotherham (2005a, b), but the potentially problematic nature of their naturalisation was not really recognised until the early 1970s.

Setting the Scene—Ideas of Recombinant Ecology

A major step in developing ideas of the processes and products of ecological recombination was a rather modest but landmark conference in the UK on 13th July 1999 (Barker 2000). Organised by George Barker, the pioneering and globetrotting urban specialist for the government's conservation agency, English Nature, and funded by the Natural Environment Research Council's URGENT programme, the event brought together leading practitioners and researchers. The intention was to find out what we know about these newly emerging, largely urban communities, what more we needed to know, and how this awareness might feed into future conservation. Held as part of the UK Forum of the Man and the Biosphere Committees, the meeting produced a modest set of published proceedings, and largely, for a variety of reasons, the initiative founded. Urban ecology certainly continued, but mostly this opportunity to grasp the nettle of recombinant communities was lost.

However, the notes of the meeting still provide a good, if often overlooked, starting point for a discussion of recombinant ecology. Interestingly, in the introduction to the proceedings, the presence of recombinant communities in rural as well as urban settings is noted. It was also stated that the term recombinant was taken from Soulé (1990) and was understood to mean the following:

'... the ecology of communities of plants and animals, the constituent members of which are drawn from a wide range of global biogeographic zones. These communities typically are mixtures of species long established in the wild in the area concerned and species relatively recently established there in the wild. In much of the literature, these are referred to respectively as 'natives' and 'aliens'. The cut-off between the two is defined differently in different places and from one piece of literature to another.'

There is a further point of note:

'... the continental European convention is taken where newcomers are defined as species established since 1500AD (roughly the point at which European world-trading accelerated) although some, but not all, species established earlier as a consequence of human activity are also included in the literature and in everyday thought as 'aliens'.'

Barker also highlighted the issue that ecological science dispassionately describes and observes the processes, and defines species interactions, whereas nature conservation introduces value judgements and management practice. The meeting sought to identify what an understanding of these processes and the recombinant ecologies might mean for British nature conservation philosophy and practice. One important suggestion from experts in the field at the time was '... that change in the species composition of ecotopes is continuous, unpredictable in direction and, effectively beyond human control.' Consequently, '... this has implications for programmes which seek to maintain ecotopes developed in the past; to restore selected species locally extinct in the recent past; to prevent the local loss of any more species; and to prevent local gain of species not currently, or only recently, established there in the wild.'

These observations remain as significant today as they were at the time. However, the relevance of such insights for urban ecology in particular, for wider landscapes more generally, and for conservation practice, has been largely ignored. There is a further point relating to the balance of local extinctions and replacement by alien, exotic species. This is that there is a loss of locally distinctive species worldwide but local diversification by invaders. However, the latter will not compensate for the former, which amounts to global biodiversity loss. (The ideas of Thomas (2013) on biodiversity response to climate change should be noted here, though the suggestion of species increases may be wildly optimistic.) Nevertheless, the idea often put forward that there will be associated loss of ecosystem functionality and ecosystem services through reduced biodiversity due to invasions is generally unsupported by any evidence. The losses of 'native' species may be undesirable from a human nature conservation perspective, and that value judgment may be the strongest argument to try to save them.

In addressing the role of recombinance in the changing nature of global ecology, it is important to set out the basic concepts and mode of arrival then invasion are significant. Yet the ecological effects, beyond the immediacy of invasion and species displacement are rarely considered, though with the work of Gilbert (1989, 1992a, b), and of Barker (2000), this began to change. Across the globe, communities are becoming urbanised and the rate of urbanisation is growing. Indeed, since the early 2000s, over half the world's human people have been urban. This is triggering issues of cultural severance in rural areas that remain, but also other issues for ecology and ecosystems as they respond to change.

With global transport and communications, there is a 'Disneyfication' of ecology with species transported across and around the planet. However, today it is mostly in and around the cities and towns that the processes of introduction, mixing and invasion take precedence. It is this ecological mixing that I refer to as 'eco-fusion' and it generates new hybrid or recombinant ecologies. Importantly, it should not be assumed that these processes are either new, or essentially urban. Recombination of ecology has been an on-going phenomenon ever since people themselves colonised and manipulated the landscape. Only now, in the age of the Anthropocene (Steffen et al. 2007), the scale of human domination over nature and with the advent of a truly global, urban society has the process come to the fore. This book takes a largely British case study approach to an important, emerging paradigm in global ecology.

The first chapter sets the scene on the idea and the need for recombinant ecology as a concept. It begins with the origins of the term and its application by researchers and scientists in the 1990s and early 2000s, relating to an overview of the landmark English Nature/MAB workshop in July 2000 (Barker 2000). This event brought together a small group of urban ecologists to consider the need for new approaches to the paradigms of urban ecosystems. The ideas of recombinant ecology were also highlighted by Rotherham (2005b) and discussed further in Rotherham (2011).

Then, as discussed in Rotherham (2011), taking the work of leading ecologists from the 1930s onwards, the way in which contemporary ecology has framed the issues of the urban and the invasive or exotic is explored. From Charles Elton's seminal volumes on animal and plant invaders (Elton 1927, 1958, 1966), to Oliver Gilbert's 'Ecology of Urban Habitats' (1989), the evolution of invasion biology and of urban ecology are considered. In the present account, reference is made to key papers and other writing from around the world as the study of invasions and of invaders has grown. Recent trends of horticultural approaches to urbanising landscapes, at odds with urban conservation in the 1980s and 1990s, are considered, (for detail on this see for example, Rotherham 2015b). In particular, the tensions between our desires to conserve the 'native' and to eradicate or at least control the 'alien' are discussed and evaluated; but so too is a current approach that seeks to relegate the native in urban environments. Ecological recombination implies recognition and acceptance of a new synthesis of native and exotic species to form novel ecologies. This does not mean there is no need to manage problem species, but merely that we should do so with a clear understanding of why, how, and indeed what the consequences may might be. There is also the uncomfortable inevitability of some of the changes now taking place.

Alongside naturally occurring processes, hybridisation of ecology has been driven by long-term nature-human interactions. These occur in agriculture and forestry, and increasingly occur through urbanisation and related environmental change (Freedman 1995). Changing ecology and ecosystems are a result of urbanisation, globalisation, climate change, and human cultural influences and the consequences are subject to ongoing research and debate (e.g. Johnson 2010; Hobbs et al. 2013; Prins and Gordon 2014, and many others). Accelerating globalisation, and both human-induced and natural climate change, speed the hybridisation process. Human impacts include disturbance, nutrient enrichment, habitat replacement (formation and destruction), and planetary-scale species dispersal

(Rotherham 2014a; Douglas et al. 2011). However, the ecological processes driving these changes are the 'natural' mechanisms of ecological succession and change, with consequent hybridisation and adaptation of species and ecosystems.

There have been attempts to identify and define the distinctive 'urban' ecologies and urban species, and this has led to the idea of 'synurbic' ecologies (Francis and Chadwick 2012). Whilst this term has been used to refer to species colonising or found in urban ecosystems, Francis, and Chadwick consider that too simplistic. They suggest the term be reserved for species populations with higher densities in urban areas than rural ones to provide a quantifiable measure of urban association. Their 2012 paper usefully clarifies and defines 'synurbic' and 'synurbization' and then addresses the issues of defining 'urban'. They provide some detail of the positive responses of 'urban' species that may generate synurbic populations. In this context, they raise the issue of whether responses drive directional selection leading to adaptation and genetic differentiation occur within the normal range of phenotypic plasticity. In terms of species responses to urban environments (such as frequency and significance of adaptation), they note the importance of understanding how synurbic populations emerge. This knowledge has implications for urban biodiversity and future management. However, as explained later, the urban view is just one, albeit important, aspect of emerging recombinant ecology. Furthermore, the urban landscape often includes strongly synurbic communities and abundant alien species, but also 'encapsulated countryside' with ecologies descended from 'ancient, traditional landscapes'. The latter are frequently land-locked in the urban environment and suffer cultural severance because of the ending of traditional and customary management. There is some resonance here with the work of Hobbs et al. (2013) and of Meurk (2011) on historical ecosystems and changes through human impacts. However, the literature on these encapsulated but ancient ecologies is relatively sparse but with one exception being the study of old-growth urban forests by Loeb (2011).

Today a key factor is that species are mixing at a rate unprecedented in the history of biodiversity evolution and this is generating obviously novel ecologies (Hobbs et al. 2013; Jørgensen et al. 2013). Now recognised as the 'Anthropocene' (Steffen et al. 2007), in a new evolutionary epoch, nature is adapting to this new canvas and altered template. A major result is the advent of ecological fusion or 'eco-fusion' as a dynamic, ongoing process of interactions as species 'native' or 'alien' to particular locations or regions to form newly combined ecological communities. Species may enter these novel communities whilst others, are displaced (Jørgensen et al. 2013; Hobbs et al. 2013). The '*Anthropocene*', is typified by human-driven influences to which nature responds (Steffen et al. 2007; Rotherham 2014a). Given this scenario, it is increasingly important to understand environmental history in order to provide context for mainstream ecology. This helps to further understanding of the drivers of these changes and to improve the predictability of future ecological outcomes (Hall 2009; Jørgensen et al. 2013; Rotherham 2014a, b; Smout 2000). The need to understand better, the evolving ecologies and thereby inform planning processes has been growing over recent decades (e.g. Douglas et al. 2011; Forman 2014; Hough 1995; Sukopp et al. 1995).

This need is particularly the case as urban areas spread and the remaining rural zones are further disrupted by human activities and by climate change.

The wider context of these changes is one of critical aspects of ecology and ecosystems under stress at every level from local, parochial, to global and planetary (Adams 2003; Barker et al. 1994; Gaston 2011; Rotherham 2014a). These processes are driven by globalisation, climate change, urbanisation, and other human cultural influences (Niemelä 2011; Sukopp and Hejny 1990; Agnoletti 2006; Agnoletti et al. 2007). Importantly, many changes are predictable through the application of informed knowledge of ecosystems and of species strategies (e.g. Grime et al. 2007; Hodgson 1986; Rotherham 2014a).

Biodiversity has become part of the critical mantra of nature conservation, and in its broadest sense, though frequently only poorly defined, refers to ecological diversity. Ecological diversity reflects underlying biological and ecological processes around the world. For a particular locale, for example Britain, diversity results from matrices of geographical spaces (habitats with parochial environmental conditions with diverse stages and states of flux and stability), and total 'biodiversity' is a summation of this. With innumerable sites, both species-rich, species-poor, this forms the national ecology. Importantly though, it is not static, but a shifting, drifting, fluxing, human-influenced, nature-influenced, climate-influenced, resource. In a context of time and space, this resource is not a fixed or finite entity. Furthermore, over immensely long timescales, evolutionary processes generate new species and drive others to extinction. Geological forces cause massive movements of continents with extinctions and associated phases of rapid evolution. In lesser timescales, periods of glaciations, inter-glacial episodes, and ice ages, stress and test ecological systems. All the other changes happen against this planetary backdrop, a 'broader canvas' of dynamic, shifting ecologies (Rotherham 2014b). To some extent these issues have been raised by Grime and colleagues (e.g. Grime 2005), in suggesting that we should not single out invasive aliens as something different and distinctive but merely another suite of invasive species behaving as invaders, alien or native to a locale. Indeed, it is also suggested (Grime 2005), that alien invasions are played out against a backdrop of much greater changes in land-use and in climate change. Grime (2002, 2003), provides in-depth discussion of some of these trends and their drivers.

Novel Ecosystems and Recombinant Communities

The best current account of novel ecosystems and the conceptual background to the ideas associated with recombinant ecology is undoubtedly the volume written and edited by Hobbs et al. (2013). Interestingly, however, they appear to overlook totally work on recombinant ecology in Britain and in Europe. The authors provide an excellent overview of the emergence of the novel ecosystem concept in relation to the history of academic ecology, and particularly the long-running debates around species, communities and classification. The present volume does not allow for a lengthy discussion of such as fundamental conceptual paradigm. In short, the debate is based around the question of whether plants and animals function as identifiable and classifiable 'communities' or merely as groups of organisms temporarily occupying the same spatial and ecological envelope. They argue that the latter idea, the 'individualistic' concept, provides the basic material from which their 'novel' ecosystems are constructed i.e. species fluxing in time and space through ecological trajectories driven by environmental forces such as climate. In this way, communities are not static, constant entities but subject to perpetual flux, dependent on their environmental drivers. One complicating factor of course, is that not only does the environment influence the species and hence their communities, but they in turn influence each other and the underlying environmental condition. Essentially, species move independently of one another in relation to environmental changes and gradients, but at the same time, they influence each other and the environment. The shifts may be rapid and repeatable as seen in secondary successions, or longer-term and directional in primary successions and when base-line conditions (like climate) change.

Hobbs et al. (2013) bring another key factor into play as they develop their model concept, human impact. They note that people generate massive, directional, permanent perturbations in ecosystems and the environment and these are hugely important in creating the conditions for novel ecosystems. However, Hobbs et al. whilst noting human impact do not address the important roles of cultural and traditional impacts over millennia, often in generating particular ecological communities and especially in then maintaining them. Furthermore, a corollary of these less obvious impacts is the effect, especially in recent decades, of their abandonment i.e. 'cultural severance' (Rotherham 2008, 2009). This is discussed in more detail later.

Hobbs and colleagues focus strongly on the critical disturbance of the environment by human agency, noting earlier work by Milton (2003) suggesting that an emerging ecosystem was one whose species composition and relative abundance had not previously occurred within a given biome. Studies at this time encompassed examples from around the globe with the impacts of human disturbances, increases in exotic species, and sometimes the displacement or collapse of 'historical ecosystems'. Hobbs et al. (2006) went on to define novel ecosystems as ones which:

(1) Demonstrated novelty with new species combinations with the potential for changes in ecosystem functioning.
(2) Human agency in terms of deliberate or inadvertent action but not depending on human intervention for their maintenance.

The second of these characteristics of a novel ecosystem would actually rule out from the definition many human-generated landscapes and ecological systems maintained through long-term customary or traditional management. They considered further the point at which human interventions can lead to novelty, and then the stage at which novel ecology becomes a 'novel' ecosystem, and finally, what happens in terms of human agency once a novel system is established. Another point made was that with more supposedly novel ecosystems on the rise but historical ecological systems on the decline, an understanding of the functioning of these new ecologies was long overdue. Hobbs et al. noted that human impacts may be 'on-site' such as direct disturbance or species introductions, or 'off-site', like climate change, nitrogen deposition etc. They discussed the pervasive nature of human impacts but also tried differentiate from a concept that since all ecosystems are modified they are therefore 'novel', to a system that is more pragmatic and useful. In addressing these issues, they noted that some human influences are deliberate and others accidental.

In order to draw out useful strands of the novel ecosystem hypothesis, Hobbs et al. (2006) placed their perceived ecosystems on a single, linear axis spanning from 'wild/historical' to 'intensive agriculture'. This was based on the idea of agricultural and other intensively managed areas being highly modified and driven by human agency until abandoned, and 'wild' or 'historical' ecosystems becoming novel as invasive species colonise triggered by human agency. Agricultural and other intensively managed ecosystems can become 'novel' as abandonment leads to new colonisation and establishment of species. They make the point that for the development of 'novel ecosystems' humans do not prescribe the outcomes as they do in agriculture or forestry, but rather the ecology responds to human influences to generate its own outcomes (Fig. 1.1).

They also place novel and hybrid ecosystems into a concept of linear placement along duel axes of biotic composition and abiotic conditions and use this to assess whether ecosystems have developed sufficient divergence from their origins to be distinctively novel ecosystems. The threshold point is where the system has moved so far that it cannot return to its starting condition. In this respect, they separate

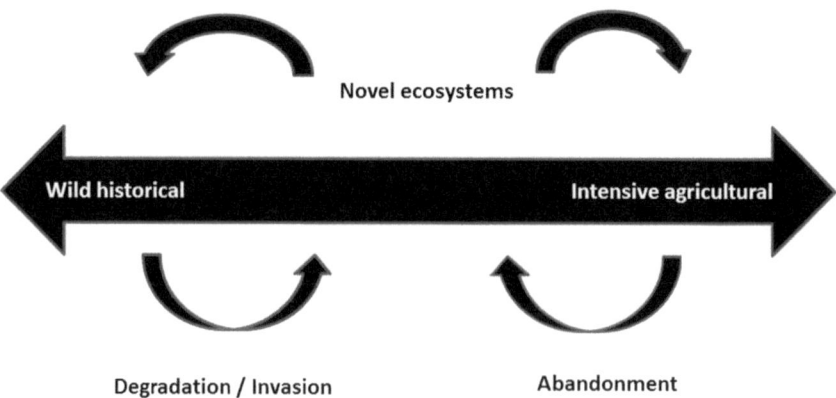

Novel ecosystems arising from degradation and invasion of wild or natural / semi-natural ecosystems, or from the abandonment of intensively managed ones (From Hobbs et al. 2013)

Fig. 1.1 The scope of novel ecosystems

novelty and hybrid systems from their concept of 'novel ecosystems'. The changes are not reversible but there may be further fluxes within the new systems. Essentially, the shifts seen in the development of hybrid ecosystems are felt to be potentially reversible to their original condition (Fig. 1.2).

Of course, there remains an issue about how far the ecology has travelled and how reversible or not the system might be. With human agency acting as an originating but not maintaining driver in truly novel ecosystems, the distance from original biotic and abiotic condition is important. The system has to have crossed a critical threshold beyond which a return to original condition, biotic or abiotic, is not possible. Hobbs et al. also postulate a conceptual framework for novel ecosystems, which relates to their origins, to human interventions like restoration, and to their degree of abiotic or biotic novelty (from Hobbs et al. 2009, 2013) (Fig. 1.3).

There are some issues with the framework and the definitions since the 'historical' ecosystems are defined as never being altered by human agency, whereas as discussed elsewhere (e.g. Rotherham 2014b), even apparently pristine ecosystems such as the Amazon rainforest for example, is often deeply affected by human cultural usage. At the other extreme, the 'used' and 'restored' ecosystems exhibit very high levels of human impact or 'design'. Restoration may seek to manipulate abiotic or biotic conditions to mimic a perceived historical ecosystem, and agriculture has human utilisation to deliver essential products. The latter is generally related to food production, but increasingly too, to chemical and energy harvesting.

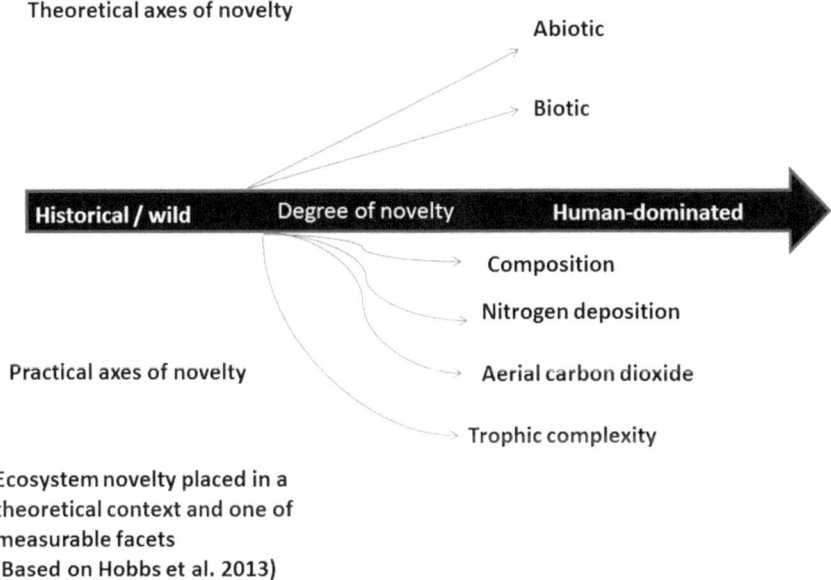

Fig. 1.2 The historical/hybrid/novel continuum

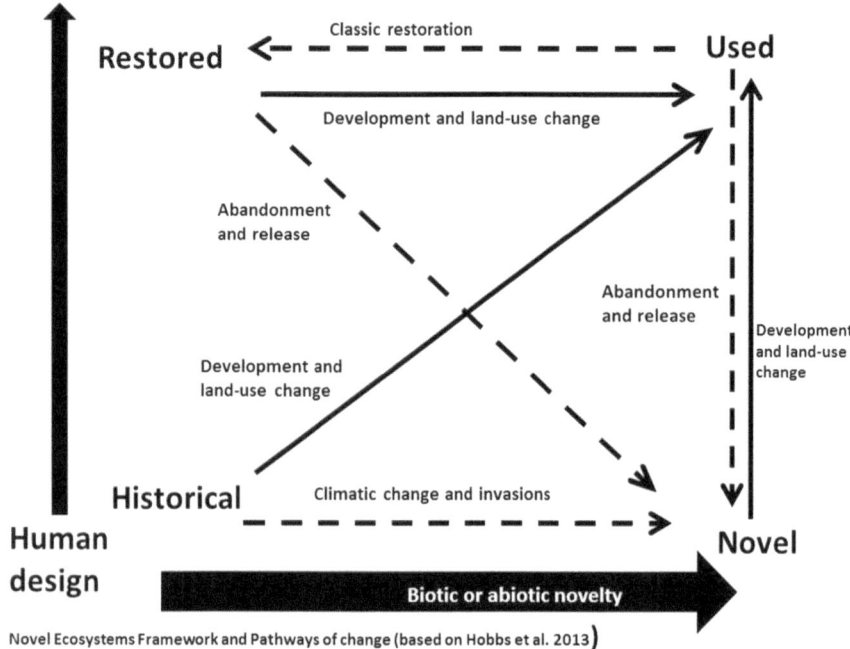

Novel Ecosystems Framework and Pathways of change (based on Hobbs et al. 2013)

Fig. 1.3 A conceptual framework for novel and hybrid ecosystems (after Hobbs et al. 2013)

Though this may vary through time and with context, the framework allows for the degree of novelty or of human design to be quantified. This is a useful conceptual development though neither precise nor rigid. Mascaro et al. in Hobbs et al. (2013) conclude their overview of novel ecosystem origins with some useful summary points as modified below:

(1) A novel ecosystem IS NOT one that is the same as would have occupied a particular space historically.

(2) It is NOT ONE that is managed intensively for specific production or which is built over.

(3) It is NOT managed in order to reproduce the historical ecosystem—i.e. classic restoration.

(4) A novel ecosystem IS a mix of abiotic, biotic and social components and interactions, which, through human influence, differ from the 'historical ecosystem'.

(5) A novel ecosystem tends to self-organise and display novel characteristics in the absence of intensive human intervention.

(6) Hybrid ecosystems are differentiated from novel ones by practical limitations in terms of ecological, environmental, and social thresholds, in relation to any potential recovery to the historical condition.

In many situations in which the landscape and its ecology are biocultural or eco-cultural in nature and origin, some of the above may be problematic (Agnoletti and Rotherham 2015). However, the idea of novelty provides a useful framework for discussion, even though the total separation of people from nature may be to misunderstand landscape history (Fig. 1.4).

Whether the various states of ecological condition occur as discrete step-changes or along a continuum of gradual change is still debateable. Furthermore, as Hobbs et al. argue, the discussion is important because it may guide any thoughts on future management, on goals and objectives for novel ecosystems, and on the dynamics of future ecosystem change. The argument is also that identifying an ecosystem condition as novel may lead to the jettisoning of aspirations to manage a site back to historical condition—so an important practitioner issue.

The debate accepts that many situations have novel elements in hybrid ecosystems that can be managed back towards the historical condition, and that this may be a valid option. It is noted however, that the combination of a consumer-driven society and the concepts of ecosystem service delivery rather than conservation, may be problematic—so anything goes as long as services are provided. Hobbs et al. have coined the idea of 'hybrid systems' to cover the situation where there may be novel elements such as invasive exotic species and altered soil

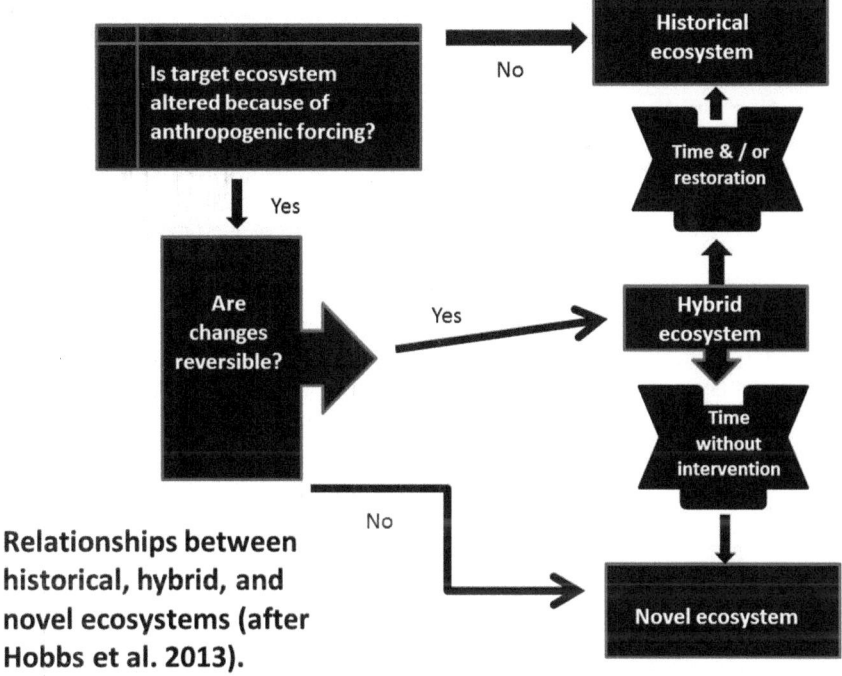

Relationships between historical, hybrid, and novel ecosystems (after Hobbs et al. 2013).

Fig. 1.4 A simple diagram to illustrate the relationships between historical, hybrid, and novel ecosystems (after Hobbs et al. 2013)

conditions, but a return to historical condition is possible. However, progress towards restoring functioning historical systems may be hindered by difficult environmental constraints and by social, economic and political barriers. In such cases, the drift may be towards the more resilient novel systems and away from the historical ones, at least in some situations.

Novelty and Recombinance

A common question asked, is the difference between a novel ecosystem and a recombinant one, or at least, a recombinant community. From the work of Hobbs and colleagues, we have a working definition of novel systems and a separation from historical and hybrid ones. Barker (2000) provided a loose but broad working definition of recombinant ecology and this with the work of Barker and of Peter Shepherd and Oliver Gilbert for example (in Barker 2000) gives an approach that works reasonably well in Britain and Europe. There seems to be a distinct separation of dialogue and approaches between the European and Antipodean schools. Meurk (2011) in Douglas et al. (2011) gives robust account of recombinant ecology though from a strongly urban perspective. He begins with a definition that 'Recombinant ecosystems comprise novel plant and animal associations that have been induced or created by people deliberately, inadvertently or indirectly. They are generally made up of indigenous and exotic species, but they may also involve associations of indigenous species alone, never before seen in nature, for example plant signature (Robinson 1993), native landscape garden designs and pictorial meadows (Dunnett and Hitchmough 2004), indigenous feature species introduced to areas beyond their natural range, or back-filled 'gaps' created by local extinctions.' Meurk notes the importance of human action, of introduced species, and of environmental change. In urban environments, many of the adapted species cope well with high levels of disturbance and sometimes stress, although the relict and often-persistent species from the historic landscape are not discussed. In the UK context, these are frequently a strong element of the urban matrix. The need to understand better the ecological envelopes of native or indigenous species entering into novel ecosystems is noted by Meurk as an important area for future research. Indeed, the work of Grime et al. (2007) in placing species into a spatial frame of stress-disturbance-competition is a useful tool in such predictions. Along with this approach to recombinant communities, it is useful to consider origins, relative proportions of native and exotic species, and issues of stability and sustainability in the face of successional changes. These factors may influence thoughts of management intervention (if any), or the likely outcomes for the ecology if these are transient or ephemeral stages. Meurk (2011) provides a simple table to aid the description of his recombinant ecosystems. Naturally, his vision is influenced strongly by experience in New Zealand where European imports have wreaked havoc with native biota for two centuries or more. Not all of this transfers easily to the European or British situation, and the definition of community, ecology, ecosystem etc., may also be problematic. Are we talking about novel elements and communities of local distribution and occurrence in a wider ecosystem context, rather than an 'ecosystem' per se? Indeed, in many British and European situations, for many communities of plants, animals and fungi, it is difficult to recognise them as functioning 'ecosystems' as such. Today, following centuries of fragmentation and transformation, many biological systems are essentially habitat patches in a sea of either agro-industrial farming or urban sprawl. Loss of ecosystem function and associated services so widely reported, are an obvious consequence of this situation. This means that we may need to distinguish between 'novel ecologies' and 'novel ecosystems', and in Britain the former predominate.

However, in the urban context Meurk raises many points of direct, pragmatic relevance to conservation planning and biodiversity conservation. He identifies a number of different and distinctive types of recombinant communities:

(1) **Remnants**—including primary and secondary vegetation—with semi-natural, pristine, degraded, recovered or additive elements, and with a lack of cultivation of the soil or what Meurk describes as non-artificial substrate. The latter is combined with a genetic link between species present in the 'primordial' vegetation and the present-day communities.

(2) **Spontaneous recombinants**—these arise from natural colonisation of disturbed ground or of vegetation, described by Meurk as 'primary' and 'secondary', and generally artificial surfaces. He also includes in this category, surrogates of naturally disturbed ecosystems alongside totally novel conditions, and invasion by species is noted as a spontaneous additive condition. Meurk includes some planted species amongst the invasives and notes that they may form 'basins of stability' and either inhibit or 'stall' ecological successions.

(3) **Deliberative recombinants**—may include deliberately over-sown mixes onto cultivated ground and designed, landscaped or planted as gardens, 'plant signatures' or restoration projects. Here he includes designer communities such as 'pictorial meadows' (e.g. Dunnett and Hitchmough 2004), and presumably some of the mixes used by organisations such as Landlife that rely on nationally, if not locally, natives.

In terms of spontaneous recombinants, Meurk notes what he describes as the 'more sinister examples of spontaneous spread of alien species', the importance of wastelands and built structures such as pavements. Presumably, this grouping would also include walls, other built or formerly built structures, and both post-industrial and post-housing areas. In these situations, there may be very distinctive and sometimes persistent recombinant plant communities.

Meurk discusses 'restorative activities' such as the control or removal of alien invasive elements or pests from remnant or secondary vegetation, or the creation of new native communities by total remove of existing 'degraded' vegetation. The aim of many such restoration projects is to artificially maintain the community until it becomes self-maintaining or at least self-regulating. However, Meurk observes that periodic intervention is usually required in order to prevent 'pests' from taking over. Furthermore, and a point I have made elsewhere, (e.g. Rotherham 2011), is that whilst restoration may be important in compensatory ecology, it does not necessary produce something that is authentically the same as the original land-scape. This is especially so when the earlier ecosystems were significantly eco-cultural, such as the English fenlands, and the new ecology, even when pop-ulated by native species, will be substantially a novel ecosystem. Meurk implies the need for an accommodation in future landscape visions for 'reconciliation ecology' to take into account indigenous and non-native species in novel combinations. However, he is writing from a New Zealand perspective where indigenous species have been spectacularly altered, firstly by Maori colonisation with associated extinctions, and then by more recent European arrivals. Quite reasonably, he worries about the negative impacts of aliens but accepts the inevitable hybridisation of the new, emerging ecologies.

Mulcock and Trigger (2008) have argued that the term 'natural' is subjective and a cultural issue and value, but the word 'native' is a factual statement. However, they note as well, that some argue that over time, 'alien' species are converted to 'native'. Meurk raises the question of whether hybrid landscapes and hybrid ecosystems or communities can be simply dismissed merely because they are perceived as not 'native' or 'natural'. He goes on to note that 'There is virtually no pristine environment in the world, especially in the predominantly biologically recipient nations; all are hybrid'. Additionally, whilst observing that such evolving systems should not be dismissed, he suggests that there are intrinsic, moral, and legal reasons why the inexorable displacement of native species by exotics is not acceptable. Meurk concludes that we have no choice in that the ecology of the world has and in future will increasingly become, recombinant. From this important conclusion, he raises the question of whether these novel systems will retain or develop local and regional character and distinctiveness rather than acquiring a globalised homogeneity. He identifies values in recombinant ecology in what he describes as 'eco-revelation' as a means to 'read' the landscape and this sits alongside the same process for long-established or native communities. Meurk states that recombinant ecologies have intrinsic interest and may help maintain biodiversity through what Rosenzweig (2003), Miller (2006) described as 'managed coexistence' or 'reconciliation'. Again though, Meurk notes that this should not be seen as dismissing the need for the protection of 'natural' or 'pristine' environments and landscapes wherever possible. He advocates what he describes as a 'gradient management approach' that would include preservation of semi-natural commu-nities through established nature conservation practice, and the intensive manage-ment of certain recombinant biotopes that support some indigenous species. The approach is justified by recognition that all combinations of species on the planet provide information and contributions which help define ecological niches as the basic building blocks of ecosystems. As such, the diversity of ecological mixes helps in the accommodation of a variety of social, cultural and 'natural' values. An interesting point raised by Miller is the idea of the 'extinction of experience' whereby a changed baseline of ecology alters people's knowledge and experience and hence expectations. This is something that I touch on in 'Eco-history' (Rotherham 2014a) in relation to industrial pollution and post-industrial degradation; communities and individuals become acclimatised to accepting the unacceptable. However, as Rob Francis has suggested (pers. comm.) such changes in social awareness, memory and acceptance, may be positive for alien species and hybrid ecologies as people grow to accept them. This has certainly been the case in the past in Britain as medieval introductions of plants are now termed 'archaeophytes' and in effect are 'honorary natives'.

Meurk moves on to discuss the need for a change in research paradigms to accommodate a new nature and a vision that unites 'natural' and human cultural aspects of landscape and ecology. He observes that this change is at odds, in particular, with the established separation in former colonial countries of indigenous nature and imperial culture. Yet in the long-established landscapes of Europe, the accommodation between nature and people over millennia was achieved in areas of 'classical countryside', or what I might describe as traditionally managed, eco-cultural landscapes. Meurk comments on the fact that modern agri-industrial growth has swept away or at least radically modified the ecologies and caused recent but probably long-term regressions in character. He attributes the drivers of these changes to the switch to single value approaches to countryside and to the need for short-term economies of scale of a capitalist approach to resource utilisation.

When European colonists spread around the world, they took with them species of animals and plants from home. What is more, the settlers quickly sent back fauna and flora from the New World to the Old. The imports and exports of species have always included both deliberate and accidental, as fauna, flora, fungi and the rest hitch a lift to lands of new opportunities. Meurk makes the prediction that '... nature will follow the model of European cultural landscapes; not an exact replica of Europe but regional representations of the structure of English countryside and urban vegetation with increasing indigenous composition'. Furthermore, he adds, 'Here will be found 'natural' patterns and ecological meaning, as much as in the remote alpine wilderness', though he skirts the issue of a leap from cultural to agri-industrial landscapes and of the potential consequences of such a shift.

Meurk closes with the thought that we cannot simply wish away recombinant ecologies and so every country and human community will have to make decisions and take steps to manage the consequent problems. He suggests that this approach will require:

- Purpose
- Goals
- Planning policies and principles
- Strategies
- Frameworks
- Methods
- Menus

- Designs
- Criteria for standards
- Measurable outcomes
- Information.

These will help guide responses and to inform the wider community.

One factor which influences the outcomes of plant introductions to new territories is the previous removal of animals, and particularly, grazing herbivores. He notes this as a particular issue in the New World, though in practice it has been a major influence globally and indeed, over millennia. We should also add to this mix, the reintroduction of one-time native animals, and the impacts of new introductions.

As noted by Meurk, the idea of strategies of disturbance, competition, and stress [as presented by Grime et al. (2007)], provides a very helpful tool for predicting outcomes and for guiding any interventions to manipulate interactions. He then suggests that in applying these key principles in restoration ecology, and by applying social considerations too, it is possible to arrive at the most effective ways in which to maintain 'natural' environments in a human context. Meurk and indeed Holmes (2005) note that 'urban wild' or 'suburban safari', dominated by exotic species, may tip the balance with spontaneous communities in many areas regardless of conservation interventions. Gilbert (1989) felt that these systems would gradually progress along successional pathways so that remnants and recombinants converge towards a Clementian mature or quasi-stable state. In terms of vegetation, Meurk predicts that some planted elements may retain dominance for long periods, whilst other communities might be diverted or move along different trajectories. New species combinations depending on particular sets of conditions may trigger novel trajectories of successional processes leading to varying states of stability.

In this context, Hulme et al. (2008) and, Meurk (2011) describe how, with a context of increasing globalisation and homogenisation, future regional ecologies will mix introduced and indigenous or aliens and natives. Increased transport of people and goods around the planet, with growing uniformity of cash-crop produce and with large-scale industrialised forestry and farming, and with rapidly expanding urbanisation, the trend is inevitable. Despite directives and mechanisms such as the Convention on Biological Diversity for example, which provides an imperative for nations to maintain their own distinct genetic resources, the powerful drivers are globalising economics and cultures. This provides an inexorable squeeze on local, regional, and national distinction and ecology but the processes disadvantage less common endemics of globally isolated communities. As species are moved and mixed, smaller island ecologies are frequently vulnerable. Continental ecologies and those of larger islands close to them, generally have large gene pools and narrow ecological niche spaces resulting from long-term jostling for space and consequent adaptation. These environments are difficult for species from smaller gene pools to penetrate; they are highly competitive. Smaller, isolated, ecologies on the other hand, present the opposite situations and are extremely vulnerable. Considered from a long-term, evolutionary perspective, this might be a reckoning up of genetic adaptation and competition as isolated faunas and floras have spawned species to fill a diversity of vacant niches. A Galapagos finch for example, over time in the absence of a woodpecker, can evolve a new species to fulfil the 'woodpecker' role. However, if faced by a real woodpecker that evolved in the harsh competition of a continental gene pool, the pseudo-woodpecker fails and becomes extinct. As globalisation breaks down the barriers, such extinctions happen repeatedly. Continental ecologies are themselves vulnerable to invasions but generally from other continents and often triggered by some catastrophe or environmental stress. Grime and colleagues have considered such issues in detail in terms of resource allocation and availability in the ecosystems (e.g. Grime 1986; Thompson et al. 1995; Burke and Grime 1996; Davis et al. 2001), and these ideas were summarised in the Report of the Unit of Comparative Plant Ecology, 1999–2003 (Anon 2004). This presented 'a general theory of invasibility' and highlighted the suggestion that much can be learned about invaders from knowledge already gleaned about colonisation processes with native species. Their long-term experimental works focussed on disturbance regimes and mineral additions or availability, and indicated an intermittent nature to the invisibility of particular communities or ecosystems. Work with Mark Davis then led to a theory, which proposed that invasibility increased with available resources. Such increase in gross resource supply may occur through reduced uptake by the 'resident' vegetation, or because of externally generated eutrophication.

The general theory of invisibility suggested the following key predictions and these are highly relevant to understanding the processes of recombination:

1. Environments subject to strong fluctuations in resource supply will be more susceptible to invasions than those with more constant supply rates.
2. There will be no general relationship between the average productivity of a plant community and its susceptibility to invasion.
3. There will be no consistent relationship between the diversity of a plant community and its susceptibility to invasion.
4. Invasibility will increase following disturbances, disease and pest outbreaks that cause direct release from damaged tissues and/or reduce capture by the resent vegetation.
5. Where increases in available resources are intermittent they will lead to successful invasion only when they coincide with the availability of invader propagules.

[Based on the Report of the Unit of Comparative Plant Ecology, 1999–2003]

Eco-Fusion

The process of invasion, hybridisation and recombinance occurs in many diverse forms and at different levels within the ecosystem, and over differing time-scales. The basic process of eco-fusion to generate novel ecologies is presented in Fig. 1.5. The result or outcome is a recombinant ecology.

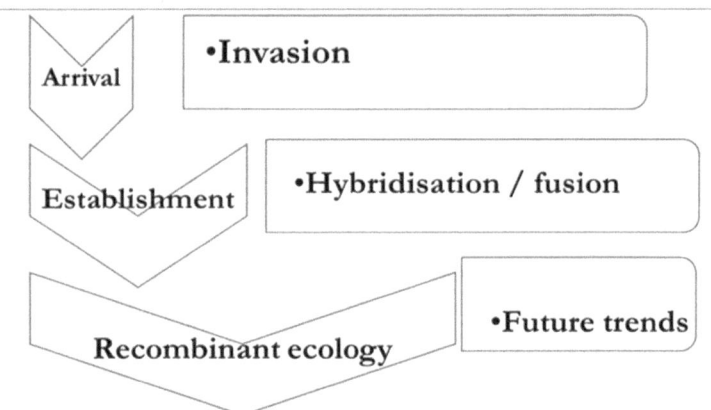

Fig. 1.5 The general process of eco-fusion

Combined with landscape-scale transformations due to human activities throughout history, the results of eco-fusion are that few ecological systems in a country such as Great Britain can be considered as predominantly 'natural'. Indeed, most landscapes and their ecological systems are radically altered, all are 'eco-cultural' (see Rotherham 2014a), and some are at most 'semi-natural'. Stand in lowland England, even in a rural area, and in most cases what you gaze out across, is an eco-cultural landscape often with exotic species and many communities altered by long-term recombinance processes. In all but a few areas that escaped parliamentary enclosures in the eighteenth and nineteenth centuries, the lowland countryside is dominated by exotic species, by human transformations, and by domesticated cultivars. Even the most apparently natural plant communities such as sand dunes, salt marshes, and clifftops, are frequently hybrid systems of long-term natives, garden escapes, agricultural and forestry species and cultivars. These may occasionally include or even be dominated by newly evolved hybrids (like the *Spartina* complex of cord grasses on salt marshes, and the hybrid or mutant *Galeobdolon*, variegated yellow archangel, in deciduous woodlands).

Then, if we turn our gaze to the upland zones, where human influence seems less, there is still an ecology altered dramatically from the 'natural' systems. Human land management over several thousand years has shaped and changed soils, drainage, and ecology. Whilst these areas retain a stronger 'natural' feel and appearance, the truth is that yet again, they are eco-cultural systems with highly modified communities. Even remote mountain areas have been altered by acid rains, by nutrient fallout from the atmosphere, and are increasingly modified by recreational disturbance and the spread of invasive alien species. Humans removed most large predators and introduced farming livestock and free-ranging herbivores like deer and the ubiquitous rabbit. All these changes have determined today's ecology. Away from the high mountains, on moor, bog, and hilltop, farmers and other land managers have for centuries, drained, burned and 'improved' the land;

once again resulting in an eco-cultural ecology. Heather moors that were diverse, relatively species-rich communities only a hundred years ago, have been transformed into *Calluna* (heather) monocultures with *Pteridium* (bracken) stands, and poor grassland of *Nardus* (mat grass), *Molinia* (purple moor-grass), and *Agrostis* (bent grass). Areas of remote and apparently 'natural' upland vegetation can be significantly dominated by the heath star-moss (*Campylopus introflexus*) an invader from Australasia first recorded in Britain in 1941; again a recombinant community.

So, even a quick review and overview suggests that much of our ecology is long-term eco-cultural, often in part at least recombinant, and subject to changes of varying scales and time-scales. It is useful to end this first chapter with some obvious examples of strongly recombinant ecological communities.

Examples of Recombinant Ecologies

It is informative to consider some very striking examples from Britain of recombinant communities and their ecologies.

Urban Riverside Vegetation

Subject to major natural disturbance with seasonal, annual cycles of flood, erosion and deposition, riverine communities offer many openings to invasive and exotic species. This is especially so in urban river corridors where major pollution and physical transformations of habitat create new opportunities as well as removing native, established ecologies. As eloquently described by Gilbert (1989, 1992c), such as communities may develop as distinctly novel, urban ecologies. Cutting across the landscape and through other habitats, they offer opportunities for species to colonise into the river corridor, and a mechanism to move downstream carried by the current. Disturbance provides space for opportunistic colonisers, both native and exotic. Particularly in urban environments Gilbert (1989, 1992c), rivers are grossly transformed by chemical and physical pollution, and by physical alterations such as canalisation and culverting. The results are radically changed environments with the suppression or removal of many natives, and major opportunities for exotic colonists. Vegetation may have natives like woodland flowers, bluebell, dog's mercury, wood sorrel, greater stitchwort, and wood anemone, mixed with exotic Japanese knotweed, Himalayan balsam, giant hogweed, Russian vine, and a canopy of alien Mediterranean fig, buddleia, and sycamore, with native alder and crack willow. In amongst this mix, as with reduced pollution levels, species have re-colonised the communities, native water voles, water shrews, and otters, mix with exotic mink, brown rat, and feral red deer. The Japanese knotweed provides a pseudo woodland canopy, which reduces summertime competitive tall herbs and allows the persistence of vernal, native woodland flowers. From these urban heartlands, invasive species spread rapidly downstream to the lower catchment and more slowly upstream too. It is not uncommon for invasive aliens such as Himalayan balsam and Japanese knotweed to be moved by human agency. Thus, otherwise restricted plants can speedily move upstream and between catchments.

Urban Commons 'Meadows'

Another recombinant feature described by Gilbert (1989, 1992a), was the 'urban common' of spontaneously developed vegetation on disturbed, urban sites. These may differ dramatically from region to region, often following by biogeographic influences but combined with socio-economic drivers such as industrial collapse, and urban turmoil or transformation. The historic origin as of urban commons sites such as post-housing or post-industrial, may determine both the dominant communities, and importantly, their persistence. The urban commons vegetation described by Gilbert in the 1980s and 1990s, suggested distinctive features associated with particular regions and cities. However, it is not clear to what extent these remain distinctive or whether they ultimately coalesce to generic, common types. He also observed the parallels between the successional processes at an urban bare ground site to those in post-glacial environments, and the important roles played by garden-escape analogues of the post-glacial floras. Many post-housing communities are spectacularly diverse during the early stages but pass quickly to species-poor tall grass and herb, with scrub and young secondary woodland. Post-industrial sites on the other hand may be extensive, and with extreme conditions of soil pH, waterlogging or drought, and generally low levels of available nutrients. The results may be long-lived persistent communities with mixed native and exotic species with considerable novelty (see Rotherham 1999; Rotherham et al. 2006a, b, 2012).

The urban commons may also reflect local horticulture and heritage through distinctive plants escaped or released from culture. These may include plants such as soapwort used for particular cleansing processes, or those brought in with imported goods such as cotton and escaping into the locally recombinant vegetation. Many species are garden plants deliberately introduced or throw-outs with domestic green waste.

Urban Water-Bodies

Urban waterbodies such as canals, ponds and lakes or reservoirs, may hold native vegetation but most frequently have communities strongly influenced by exotic, invasive species. Furthermore, the composition of these communities may be quite distinctive and may vary over time as new species become fashionable for cultivation in aquaria, in parks, and in garden ponds. These species, often chosen for features such as frost tolerance may quickly spread, establish and multiply. On some occasions, the establishment may be through accidental means such as the case in Sheffield, where in the 1980s and 1990s, the City Council's Countryside Service was creating and stocking wildlife ponds in schools and in countryside parks across the region. However, the three pond plants provided to them by an aquatic supplier as 'native oxygenating plants', were New Zealand pygmy weed (*Crassula helmsii*), *Lagarosiphon* and yellow fringed waterlily (*Nymphoides peltata*). These three exotic species formed a dramatic new community, which was both visually attractive, and a Mecca for amphibians and invertebrates such as damselflies and dragonflies. From the semi-amphibious areas around the pond, dominated by *Crassula*, to the deeper water with *Lagarosiphon*, and then water lily, these communities were distinctive and persistent. Inspections of many urban still-water sites across England suggestions that the vast majority are recombinant, many have significant exotic components, and most are of conservation value to their local area.

Along with exotic plant species that spread through ponds, lakes, canals and other waterbodies, there are numerous other escapees from captivity, including both exotic turtles and terrapins, and a variety of amphibian species. These are mostly fauna kept in collections or as pets, and which are discarded into water bodies when the owner gets bored or the animals becomes too large to keep. The result of these miscellaneous introductions is the emergence of hybrid ecologies from the vegetation, to herbivorous fauna and to higher-level carnivores. One excellent example of the dramatic impact of an exotic animal has been the spread of the North American signal crayfish throughout England, displacing the native species and now clearly a significant influence in many aquatic ecosystems.

Urban Woodlands

Depending on local topography and on history, towns and cities may have more or less 'ancient' woodland cover, and these sites have strongly native vegetation components. However, the urban woodland scene also included riverine woods, both ancient and recent secondary and spontaneously regenerated sites from abandoned farmland now urbanised, and from urban commons successions. Some of these sites have significant components of exotic woody species such as sycamore and buddleia, though because such areas receive little thought or protection, they rarely develop any significant degree of maturity. This is unfortunate since it would be interesting to observe the successional processes of *Buddleia*-dominated woodland for example.

However, it is also informative to consider the more typically 'natural' urban woodlands. Many of these one-time native coppice woods have been radically affected by histories of industrial use, for example the manufacture of charcoal for metal industries. This may include the removal of vegetation and soil from much of a site (see Rotherham 2014a, Rotherham and Doram 1992). Furthermore, when traditional and industrial utilisation ceased, often in the 1800s or 1900s, the sites were frequently re-planted as high forest with trees that were either or both exotic genotypes or species exotic to the sites. Therefore, whilst these sites have strong links to long time-lines of traditional management and native species, their modern structure and ecology are significantly altered. Furthermore, in the urban environment such woods gained exotic trees, shrubs and herbs throughout the period of the nineteenth and twentieth centuries, and up to the present time. Over this period in the urban catchment, they were subjected to firstly massive acid rain and fallout of grit from coal-burning fires, and in more recent times, nitrogen from motorcars. These are transformational impacts. Into this transformed ecology, through deliberate introduction and accidental introductions from garden escapes, bird droppings (garden berry bushes of various sorts), and planting of exotics, the twenty-first century communities are truly recombinant. Native bluebells thrive under exotic beeches (generally non-native in northern England), with planted Scots pine, European larch, hornbeam, sycamore, Norway maple, and many others. The herb layers include the garden escapee, variegated yellow archangel, Himalayan balsam, exotic hybrid *Montbretia*, Japanese knotweed, sweet cicely, and much more. Exotic trees and shrubs dispersed by berries include *Sorbus* varieties, *Ilex* cultivars, and cotoneasters. Other planted shrubs include *Rhododendron ponticum*, *Prunus laurocerasus*, *Mahonia ilex*, and others. Exotic ivies (*Hedera*) can also be significant ground cover components.

Ancient woodland sites in the wider, more rural landscape may share some of these changes and impacts, though often with less drastic consequences. However, management for game shooting interests, sometimes over a century or more, may mean associated species introduced, some already mentioned in urban woods. However, shrubs such as snowberry (*Symphoricarpus*) and bridewort (*Spiraea*) were frequent additions for game preservation. The disruption of soils in rural woods was often, but not always, less.

Plantation Forests

A more extreme impact at woodland or forest level is the planting of commercial forestry sites with exotic tree species, and usually conifers. This has often been into agricultural lands, into heaths or bogs, or into existing, broad-leaved ancient woods. The activities have transformed soils and surface water, and changed ecological systems to a forced high forest. In existing woodland sites, traditional native-species coppice was frequently converted into high forest, exotic, timber production. The impacts and conservation implications are widely known, and whilst many were very negative, other effects benefitted many wildlife species. So, for example, mature conifer plantations have become very important habitats for birds such as siskin, redpoll, crossbill, parrot crossbill, goshawk, and others. Mammals such as pine marten and red squirrel may also benefit.

However, the Victorian and more recent establishment of conifer, beech, and sycamore plantations has also unleashed a wave of naturalised conifer and other species spreading rapidly in open woodland, heaths, moors and bogs. Now fully integrated into the ecology a landscape scale, these species have helped to create newly recombinant woodlands with Scots pine (exotic in England), European larch, Sitka spruce, beech, sycamore, and others mixing with natives in new and spontaneous combinations. From soil fungal associates, to invertebrate phytophages and dependent bird species, these trees are generating long-term, novel ecosystems.

In recombinant ecologies, which species are allowed and accepted and which are not?

Having established with just a few examples briefly described, the widespread, growing occurrence of recombinant British ecologies, some key questions arise about our responses to these new systems. For example, acceptance of the inevitability of recombinant ecologies raises issues of which species and communities might be acceptable, and which might not. However, whilst science and history may inform such judgements, the decisions are essentially political in nature, subjective choices based on objective observations. There are also issues around the inevitably dynamic nature of these novel communities as local environmental conditions, ecological opportunities, and especially climate, all change and flux. In terms of the raw materials of ecology, new species and varieties of fauna and flora are constantly added to the mix and ecological fusion continues.

Meurk (2011) suggested that we need to learn to 'go with the flow' and not 'fight nature head on' since 'whether we like it or not both in cultural and wild landscapes' recombinants are here to stay. Therefore, we should try to derive something useful from the mix and from the process. He goes on to advocate the following key actions in relation to recombinant ecologies:

1. Identify and eliminate pathological species;
2. Tolerate benign species;
3. Encourage beneficial, especially indigenous species—all within their place;
4. Maintain urban legibility and visibility (the full story)—sufficient for viability and sustainability, both ecologically and culturally;
5. Maintain as many permutations as possible—along management gradients and within a stress-disturbance matrix;
6. For successional species build up populations, seed sources and seed banks, and create punctuated disturbances so that indigenous and exotic species have an equal chance at being part of the mix and passing their propagules onto the next disturbed site.

Meurk (2011) wrote especially with urban environments in mind, but the list provides some baseline ideas to set alongside those of conventional conservation wisdom. Clearly, the strategies for a response to recombinance will vary with other conservation priorities, with resources and with context, i.e. where, when, and what. In an urban situation, Meurk suggested that since we cannot second-guess what is the best or optimum environmental and management condition, we have to provide as many options as possible to ensure a high probability of something for every species. His stated aim is to allow or facilitate all these species to establish viable populations with minimal management and expending as little energy and money as possible. This seems a rather broad and undiscriminating approach at odds with some of his management suggestions. He goes on to advocate 'traditional tools' such as grazing with various species, the use of fire, application of herbicides, manual weeding, soil stripping, and in many different combinations.

This approach to urban habitats seems a world away from the Oliver Gilbert view of the 'urban commons' of letting them be, to follow the ecological successional processes, and something that accords with some at least, of the aspirations for 'new wild'. Essentially the Meurk model is one that may be heavily 'gardened' to generate endpoints acceptable to the manager. This may be appropriate in some situations but not others, and perhaps applies more to the traditional nature conservation end of the recombinant spectrum. However, there are some useful thoughts on pathogens to be controlled and benign species to be tolerated. Maybe key 'problem' invasive plants and animals could be added to Meurk's pathogens, but prefixed by 'where desirable, possible, practicable, financially viable, and appropriate'. Some of these riders are essentially subjective choices and decisions but based (hopefully) on objectively gathered information.

References

Adams W (2003) Future nature: a vision for conservation. Earthscan, London

Agnoletti M (ed) (2006) The conservation of cultural landscapes. CAB International, Wallingford

Agnoletti M, Rotherham ID (2015) Landscape and biocultural diversity. Biodivers Conserv 24:3155–3165

Agnoletti M, Anderson S, Johann E, Kulvik M, Saratsi E, Kushlin A, Mayer P, Montiel C, Parrotta J, Rotherham ID (2007) Guidelines for the implementation of social and cultural values in sustainable forest management: a scientific contribution to the implementation of MCPFE—Vienna resolution 3. IUFRO occasional paper no. 19, ISSN 1024-414X, IUFRO Headquarters, Vienna, Austria

Anon (2004) Report of the Unit of Comparative Plant Ecology 1999 – 2004. Unit of Comparative Plant Ecology, University of Sheffield, Sheffield

Barker G (ed) (2000) Ecological recombination in urban areas: implications for nature conservation. English Nature, Peterborough, pp 21–24. Barker G (ed) (2000) Ecological recombination in urban areas: implications for nature conservation: a workshop held at the centre for ecology and hydrology (Monks Wood), 13 July 2000. UK Man and Biosphere Committee, Urban Forum, English Nature, Centre for Ecology and Hydrology, Peterborough, ISBN 185716542X, 9781857165425, 25 p

Barker G, Luniak M, Trojan P, Zimny H (eds) (1994) Proceedings of the second European meeting of the international network for urban ecology. In: Memorabilia Zoologica, vol 49. Warsaw

Bridgewater PB (1990) The role of synthetic vegetation in present and future landscapes of Australia. Proc Ecol Soc Aust 16:129–134

Burke MJW, Grime JP (1996) An experimental study of plant community invisibility. Ecology 77 (930):776–790

Clement EJ, Foster MC (1994) Alien plants of the British Isles. Botanical Society of the British Isles, London

Davis MA, Thompson K, Grime JP (2001) Charles S. Elton and the dissociation of invasion ecology from the rest of ecology. Divers Distrib 7:97–102

Douglas I, Goode D, Houck MC, Wang R (eds) (2011) The Routledge handbook of urban ecology. Routledge, London

du Toit JT, Rogers KH, Biggs HC (eds) (2003) The Kruger experience: ecology and management of savanna heterogeneity. Island Press, Washington DC

Dunn ST (1905) Alien flora of Britain. West, Newman and Co., London

Dunnett N, Hitchmough J (eds) (2004) The dynamic landscape; design, ecology and management of urban planting. Spon Press, London

Eisenhauer N, Milcu A, Sabais ACW et al (2011) Plant diversity surpasses plant functional groups and plant productivity as driver of soil biota in the long term. PLoS ONE 6(1):e16055. Doi:10.1371/journal.pone.0016055

Ellis GR (1993) Aliens in the British flora. National Museum of Wales, Cardiff

Elton CS (1927) Animal ecology. Sidgwick & Jackson, London

Elton CS (1958) The ecology of invasions by animals and plants. Methuen, London

Elton CS (1966) The pattern of animal communities. Methuen, London

Fitter R (1945) London's natural history. Collins New Naturalist, London

Forman RTT (2014) Urban ecology. Science of cities. Cambridge University Press, Cambridge

Francis RA, Chadwick MA (2012) What makes a species synurbic? What makes a species synurbic? Appl Geogr 32(2):514–521

Freedman B (1995) Environmental ecology—the effects of pollution, disturbance and other stresses, 2nd edn. Academic Press, San Diego

Gaston KJ (ed) (2011) Urban ecology. Cambridge University Press, Cambridge

Gilbert OL (1989) The ecology of urban habitats. Chapman and Hall, London

Gilbert OL (1992a) The flowering of the cities: the natural flora of 'urban commons'. English Nature, Peterborough

Gilbert OL (1992b) Rooted in stone. The natural flora of urban walls. English Nature, Peterborough

Gilbert OL (1992c) The ecology of an urban river. Br Wildl 3:129–136

Gillson L (2015) Biodiversity conservation and environmental change: using palaeoecology to manage dynamic landscapes in the Anthropocene. Oxford University Press, Oxford

Gillson L, Marchant R (2014) From myopia to clarity: sharpening the focus of ecosystem management through the lens of palaeoecology. Trends Ecol Evol 29(6):317–325

Godwin H (1975) History of the British flora. A factual basis for phytogeography, 2nd edn. Cambridge University Press, Cambridge

Goldsmith E, Hildyard N (eds) (1986) Green Britain or industrial wasteland. Polity Press, Cambridge

Grime JP (1986) The circumstances and characteristics of spoil colonisation within a local flora. Philos Trans R Soc B314:637–654

Grime JP (2002) Declining plant diversity: empty niches or functional shifts? J Veg Sci 13:457–460

Grime JP (2003) Plants hold the key: ecosystems in a changing world. Biologist 50:87–91

Grime JP (2005) Alien plant invaders; threat or side issue? ECOS 2093(40):33–40

Grime JP, Pierce S (2012) The evolutionary strategies that shape ecosystems. Wiley-Blackwell, Chichester

Grime JP, Hodgson JG, Hunt R (2007) Comparative plant ecology. A Functional approach to common British species, 2nd edn. Castlepoint Press, Dalbeattie

Grindon LH (1859) Manchester flora [No publisher or location given]

Grindon LH (1864) British garden botany [No publisher or location given]

Hall M (ed) (2009) Greening history: the presence of the past in environmental restoration. Routledge Publishing, London

Harris S, Yalden DW (2008) Mammals of the British Isles: handbook, 4th edn. The Mammal Society, Southampton

Hobbs RJ et al (2006) Novel ecosystems; theoretical and management aspects of the new ecological world order. Glob Ecol Biogeogr 18:1–7

Hobbs RJ, Higgs E, Harris JA (2009) Novel ecosystems; implications for conservation and restoration. Trends Ecol Evol 24:599–605

Hobbs RJ, Higgs ES, Hall CM (eds) (2013a) Novel ecosystems. Intervening in the new ecological world order. Wiley-Blackwell, Chichester

Hobbs RJ, Higgs E, Hall CM, Bridgewater P et al (2013b) Managing the whole landscape: historical, hybrid, and novel ecosystems. Front Ecol Environ 12(10):557–564

Hodgson JG (1986) Commonness and rarity in plants with special reference to the Sheffield flora. Biol Conserv 36(3):199–252

Holmes H (2005) Suburban safari: a year on the lawn. Bloomsbury, New York

Hough M (1995) Cities and natural processes. Routledge, London

Hulme PE (2006) Beyond control: wider implications for the management of biological invasions. J Appl Ecol 43(5): 835–847

Hulme PE, Bacher S, Kenis M, Klotz S, Kühn I, Minchin D, Nentwig W (2008) Grasping at the routes of biological invasions: a framework for integrating pathways into policy. J Appl Ecol 45(2):403–414

Ingrouille M (1995) Historical ecology of the British flora. Chapman & Hall, London

Johnson S (ed) (2010) Bioinvaders. White Horse Press, Cambridge

Jørgensen D, Jørgensen FA, Pritchard SB (eds) (2013) New natures. Joining environmental history with science and technology studies. University of Pittsburgh Press, Pittsburgh

Lever C (1977) The naturalized animals of Britain and Ireland. Hutchinson & Co (Publishers) Ltd, London

Lindenmayer DB, Fischer J, Felton A, Crane M, Michael D, Macgregor C, Montague-Drake R, Manning A, Hobbs R (2008) Novel ecosystems resulting from landscape transformation create dilemmas for modern conservation practice. Conserv Lett 1(3):129–135

Loeb RE (2011) Old growth urban forests. Springer Briefs in ecology. Springer, Dordrecht

Maraun M (2012) The evolutionary strategies that shape ecosystems by J. Philip Grime and Simon Pierce. Wiley, West Sussex. Book review in Basic Appl Ecol 13(8):736

McNeeley JA (ed) (2001) The great reshuffling. Human dimensions of invasive alien species. IUCN, Gland

Meurk C (2011) Recombinant ecology of urban areas: characterisation, context, and creativity. In: Douglas I, Goode D, Houck MC, Wang R (eds) (2011) The Routledge handbook of urban ecology. Routledge, London, pp 198–220

Miller JR (2006) Restoration, reconciliation, and reconnecting with nature nearby. Biol Conserv 127(3):356–361

Milton SJ (2003) Emerging ecosystems; a washing stone for ecologists, economists, and sociologists? S Afr J Sci 99:404–406

Mulcock J, Trigger D (2008) Ecology and identity: a comparative perspective on the negotiation of 'nativeness'. In: Wylie D (ed) (2008) Toxic belonging? Identity and ecology in Southern Africa. Cambridge Scholars Publishing, Newcastle-upon-Tyne

Niemelä J (ed) (2011) Urban ecology. Oxford University Press, Oxford

Pearce F (2015) The new wild. Why invasive species will be nature's salivation. Icon Books, London

Pickett STA, White PS (1985) The ecology of natural disturbance and patch dynamics. Academic Press, New York

Preston CD, Pearman DA, Dines TD (2002) New atlas of the British and Irish flora. Oxford University Press, Oxford

Prins HHT, Gordon IJ (2014) Invasion biology and ecological theory. Insights from a continent in transformation. Cambridge University Press, Cambridge

Ratcliffe DA (1984) Post-medieval and recent changes in British vegetation; the culmination of human influence. New Phytol 98:73–100

Robinson N (1993) Place and planting design—plant signatures. Landscape 53:26–28

Rosenzweig ML (2003) Reconciliation ecology and the future of species diversity. Oryx 37 (2):194–295

Rotherham ID (1999) Urban environmental history: the importance of relict communities in urban biodiversity conservation. Pract Ecol Conserv 3(1):3–22

Rotherham ID (2005a) Alien plants and the human touch. J Pract Ecol Conserv Spec Ser 4:63–76

Rotherham ID (2005b) Invasive plants—ecology, history and perception. J Pract Ecol Conserv Spec Ser 4:52–62

Rotherham ID (2008) Lessons from the past—a case study of how upland land-use has influenced the environmental resource. Aspects Appl Biol 85:85–91

Rotherham ID (2009) The importance of cultural severance in landscape ecology research. In: Dupont A, Jacobs H (eds) (2009) Landscape ecology research trends. Nova Science Publishers Inc., USA

Rotherham ID (2011) The implications of landscape history and cultural severance in environmental restoration in England. In: Egan D, Hjerpe E, Abrams J (eds) Integrating nature and culture: the human dimensions of ecological restoration. Island Press, Washington DC, pp 277–287

Rotherham ID (2014a) Eco-history: an introduction to biodiversity and conservation. The White Horse Press, Cambridge

Rotherham ID (2014b) The call of the wild. Perceptions, history people and ecology in the emerging paradigms of wilding. ECOS 35(1):35–43

Rotherham ID (2015a) Bio-cultural heritage and biodiversity—emerging paradigms in conservation and planning. Biodivers Conserv 24:3405–3429

Rotherham ID (2015b) Times they are a changin'—recombinant ecology as an emerging paradigm. Int Urban Ecol Rev 5:1–19

Rotherham ID, Doram G (1992) A preliminary study of the vegetation of Ecclesall Woods in Relation to Former Management. Sorby Record 29:60–70

Rotherham ID, Wild M, Lunn J (2006a) Pioneer Vegetation Communities from the Coal Measures of Yorkshire, England 1. *Agrostis stolonifera-Holcus lanatus* pioneer community. Int Urban Ecol Rev 1:38–44

Rotherham ID, Wild M, Lunn J (2006b) Pioneer Vegetation Communities from the Coal Measures of Yorkshire, England 2. *Vulpia bromoides-Arenaria serpyllifolia* pioneer community. Int Urban Ecol Rev 1:45–48

Rotherham ID, Lunn J, Spode F (2012) Wildlife and coal—the nature conservation value of post-mining sites in South Yorkshire. In: Rotherham ID, Handley C (eds) (2012) Dynamic landscape restoration. Landscape archaeology and ecology special series. Papers from the Landscape Conservation Forum (1), pp 30–64

Salisbury E (1961) Weeds and aliens. Collins, London

Smout TC (2000) Nature contested: environmental history in Scotland and Northern England since 1600. Edinburgh University Press, Edinburgh

Soulé ME (1990) The onslaught of alien species, and other challenges in the coming decades. Conserv Biol 4(3):233–239

Stace CA, Crawley MJ (2015) Alien plants. HarperCollins, London

Steffen W, Crutzen PJ, McNeill JR (2007) The Anthropocene: are humans now overwhelming the great forces of nature. AMBIO 36(8):614–621

Sukopp H, Hejny S (eds) (1990) Urban ecology. Plants and plant communities in urban environments. SPB Academic Publishing bv, The Hague

Sukopp H, Numata M, Huber A (eds) (1995) Urban ecology as the basis of urban planning. SPB Academic Publishing bv, The Hague

Thomas C (2011) Britain should welcome climate refugee species. New Scientist, pp 29–30

Thomas C (2013) The Anthropocene could raise biological diversity. Nature 502:7

Thomas C et al (2004) Extinction risk from climate change. Nature 427:145–148

Thompson K, Hodgson JG, Rich TCG (1995) Native and alien invasive plants: more of the same? Ecography 18:390–402

Walker D, West RG (1970) Studies in the vegetational history of the British Isles. The University
    Press, Cambridge
Weightman G (1990) Brave new wilderness. Wildlife in Britain since the industrial revolution.
    Weidenfeld and Nicolson, London

Photograph: Himalayan balsam is a Victorian wild garden flower now thoroughly naturalised

Chapter 2
An Historical Perspective of Ecological Hybridisation

Understanding human and environmental history is at the heart of an awareness of recombinant ecology. Unfortunately, since academics often work in disciplinary silos and many rarely step beyond the comfort zone of their core field, the connections and resulting conclusions are frequently missed. It is therefore necessary and helpful in this second chapter, to explore the roles of history in the making of ecology.

The concept of 'ecological hybridisation' or 'eco-fusion' is useful, and this chapter draws examples of the hybrid nature of much ecology occurring 'naturally' but often conveniently ignored or overlooked. Ecological hybridisation is taken to include species-level genetic hybrids but importantly too, the mixing of native and exotic in novel communities and ecosystems. In this emerging concept, these ideas are embedded in a longer-term view of the evolution and agricultural and forestry systems, and the impacts that these have had on landscapes and ecology. Plant collectors, gardeners, acclimatisation societies, and horticulturalists, particularly during the nineteenth century, took species into cultivation and deliberately hybridised species, and mixed and hybridised ecologies from around the world. These processes are explained with key examples. Stace et al. (2015), in their 'Hybrid Flora of the British Isles' detailed many of the hybrid outcomes of deliberate or accidental cross-fertilisations of plants, both native and exotic.

The garden examples include the so-called 'wild rhododendron', which is in fact a mix in varying proportions of the European *Rhododendron ponticum*, and the American *R. maximum* and *R. catawbiense*. The result is a highly adaptable and invasive hybrid. Other deliberate hybrids include the invader, montbretia, *Crocosmia* x *crocosmifolia*, a deliberate hybrid of *Crocosmia aurea* and *Crocosmia pottsii*. Variegated yellow archangel, *Lamiastrum galeobdolon* var *argentatum*, is probably a naturally occurring hybrid through a doubling of its parental genes, but cultivated by Victorian gardeners for its striking variegated foliage.

However, the wider context of this issue is that many native '*species*' in Britain are in fact largely hybrids or hybrid swarms, for example, the English oak and the Sessile oak occur as hybrid swarms of *Quercus petraea* x *robur*, in many

© The Author(s) 2017   35
I.D. Rotherham, *Recombinant Ecology - A Hybrid Future?*,
SpringerBriefs in Ecology, DOI 10.1007/978-3-319-49797-6_2

woodlands. This is complicated further by a significant proportion of the genes originating from European, especially Dutch stock in the 1700s and 1800s, and presenting quite distinctive phenotypes. In wet woods and moors, most willows, *Salix* sp. are hybrid complexes, and in wet grasslands, the marsh orchids form complex swarms of inter-breeding hybrids from a number of possible parents. This simply suggests that 'nature' is not as pure as it is often portrayed (see Stace et al. 2015 for a detailed account).

The Cultural Facilitation of Ecological Invasion

Throughout history, people have interacted with nature to modify and sometimes destroy environmental resources (Rackham 1986; Rotherham 2013a, 2014a). Human management has generated identifiable and distinctive '*cultural*' landscapes, as fusions of natural and anthropogenic elements. Furthermore, many of these eco-cultural landscapes have been managed with customary and traditional mechanisms (e.g. Rotherham 2013b). In a pre-petrochemical age, such management created traditional landscapes frequently made up of diverse, species-rich habitats and maintained by long-established land-use applied continually over the years. It is argued (Rotherham 2014a), that such traditional, 'unimproved' lands provided habitats for biodiversity with direct links to habitat analogues in the ancient, primeval, 'natural' European landscape (Rotherham 2009b, 2013b, 2014a), the vision of Vera (2000). I suggest that the complex inter-relationships and ecologies of, for example, species-rich limestone grasslands, fens, peatlands, and ancient woods, result from such long-standing human-nature interactions and they are analogues of the ecological communities of the primeval landscape. This approach fits well with the ideas of Rackham (1980, 1986) for example. Furthermore, it is helpful to have these concepts in mind when we seek to understand the significance of recombinance through time.

The human influences in these ecosystems include the hybridisation of species and ecology, through 'eco-fusion'. This is most readily recognised in the world's increasingly urban environments, but it clearly occurs more widely. Furthermore, whilst recombination as a process has only been recognised in recent decades, it has occurred on varying scales since humans first influenced the landscape. Today, huge areas of land are dominated by forestry and agriculture, and in these imposed environments plants, animals, and fungi move and mix beyond natural distributions and limits. This means that old, new, native, and exotic, are intertwined in novel, recombinant communities in hybrid ecosystems. Particularly now, in the rapidly expanding urban heartlands, this new ecology, of native and alien, is locked in a perpetual and dynamic struggle for dominance with the resulting formation of novel dependencies, interactions, and communities (Rotherham 2014a; Gilbert 1989, 1992a). Recognition of these processes and their consequences challenges many current debates in conservation ecology, particularly in relation to debates on alien species. The approach generates new paradigms and matters of perceptions,

**But the processes are not new – we can see them in
landscape-scale changes over centuries**

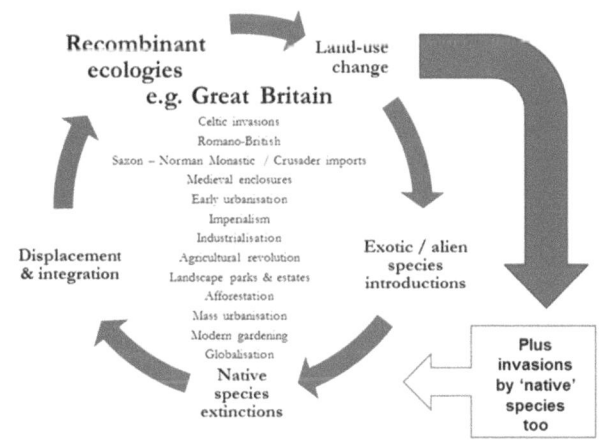

Fig. 2.1 The flow of recombinant ecology

judgements, and actions. These ideas of recombinant, hybrid ecologies and eco-fusion processes are new, and they have significant implications for future ecologies. Indeed, for discussions on ideas of 'wilding' and 're-wilding', ecological fusion and hybrid ecology, are vital conceptual frameworks for debates on future landscapes and ecologies (Rotherham 2014a, b; Prins and Gordon 2014; Taylor 2005). The results of 're-wilding' landscapes on a large scale (e.g. Monbiot 2013a; Taylor 2005; Carver 2014), will be determined by such processes acting within the contemporary environmental matrix. These areas will not reproduce some golden age past ecology, but reflecting their histories they will generate 'futurescapes' with novel hybrid ecologies. In the current debates on future 'wild' landscapes, these outcomes are neither recognised nor considered.

This historical precedence drives future change and is not new. Recombination has been a fundamental process driving and forging ecological systems for thousands of years; only today, its impacts are becoming more dominant, more obvious, and more rapid (Fig. 2.1).

Hybrid Ecologies

Over centuries, non-native, introduced species have altered landscapes and indigenous ecologies (Gilbert 1989; Rackham 1986; Rotherham and Lambert 2011a, b; Rotherham 2014a, b). Furthermore, these changes, such as by the rabbit introduced by the Normans to Britain, may be the fundamental determinants and keystone species in modern-day ecology. However, trends and changes need to be assessed in the wider context of fluxing climate, land-use, and other human-nature interactions. A brief consideration of British landscapes shows there is little genuinely pure 'natural' or 'native'. As noted earlier, most of our environment is dominated by 'cultural' or 'eco-cultural' landscapes and their ecological communities (Rotherham 2007, 2008a, 2009b, 2011, 2014a). This ecology results from centuries of interactions between people and nature, sometimes with 'semi-natural' components and some with regionally distinctive traditional countryside. However, significant proportions of today's landscapes are highly modified; large areas of British farmland for example, are of modern origins and ecologically highly exotic in their make-up.

Whilst natural changes are influential, people are at the core, directly or indirectly, of most biological invasions (Johnson 2010; Rotherham and Lambert 2011b). There are two particular examples of invasion and fusion witnessed in Great Britain and which confirm the human component of the human-nature influence. The first example is the deliberate introduction of plants and animals around the recently discovered world, by the Victorian Acclimatisation Societies. The second example is the often-overlooked, case of the Victorian Wild Garden Movement (Rotherham 2005a, b, 2014a; Rotherham and Lambert 2011a, b). In these two nineteenth-century phenomena, we see the germ of many of contemporary issues and challenges for modern nature conservation and land management. There is also an interesting twist in terms of the changing perceptions, attitudes, and politics in relation to nature and the exotic. This is often overlooked, but hugely influential from the early nineteenth century to the early twenty-first century. Davis et al. (2001) wrote of how British attitudes to exotic species changed with the seminal writings and broadcasts of Charles Elton. However, despite this, the more general and pervading influences of fashion and taste with regard to exotic species have been ignored. Indeed, the often critical roles of accidental or even deliberate introductions of now invasive species have been overlooked (see Rotherham 2001, 2005a, b, 2011) or not recognised.

Human cultural facilitation of invasions is very significant. Furthermore, perceptions and attitudes have influenced both the processes and the responses (e.g. Rotherham and Lambert 2011b). However, the research in this field is often lacking since the work inherently crosses the boundaries of ecological science and of history into a less well-trodden path of interdisciplinary study. Additionally, it is obvious that many if not most of the ecological invasions of the twentieth, and twenty-first centuries in Britain were culturally facilitated (e.g. Rotherham 2001, 2005a, b, 2009a; Rotherham and Lambert 2011b).

Furthermore, the problems associated with aggressive and invasive plants and animals are not new (consider for example, the spread of both black rats and brown rats, and their impacts on people and landscapes). Yet, the scale of impact, now combined with rapid climate change and other environmental influences, is spectacular. It is suggested that 15% of Europe's 11,000 exotic species combine to generate environmental and economic damage valued at £2 bn *per annum* to the UK economy. However, behind the often-torrid newspaper headlines remain key questions about what is native and where, what is alien, and when. This applies from Spanish bluebells, to eagle owls, Canada geese, ruddy ducks, ring-necked parakeets, Japanese knotweed, Himalayan balsam, to feral big cats, beavers and signal crayfish, and wild boar. With all these new arrivals, which ones get a '*free pass*' to be part of a new hybrid future and which do not (Rotherham 2009a, 2014b, 2016; Rotherham and Lambert 2011b)?

Examples of the Historic Cultural Drivers of Recombination

Considering the importance of history in ecological recombination, it is useful to consider very briefly some examples of the process at work.

Agriculture: Since people first settled in permanent farming communities, they have changed the landscape, the soil, the drainage, and the ecology. Furthermore, they cleared land, burnt vegetation, cultivated crops, which they modified by careful breeding and selection, and they domesticated animals that they bred into domesticated stock. These activities in turn transformed the environmental conditions and the landscape, and some species either escaped back into the wild or interbred with wild species. As humans then moved around the world, from the earliest times, they took with them their domesticated associated as a variety of other 'hangers-on' to be established into new terrain. Increasingly, especially from the Bronze Age onwards, people changed nature and this challenges many concepts of what is 'natural', native' or 'indigenous'. When you decide to 'stop the clock' becomes rather significant, since people have been mixing the species and genetic pools for countless centuries. Clearly, these impacts have grown exponentially in recent times, but the processes are ancient.

Forestry: It is important to distinguish between traditional 'woodland management' which has gone on in various forms, certainly since Roman times, but probably to some extent from much earlier still, and forestry. The latter was an invention of the French and German land managers in particular, and began in earnest during the 1700s and 1800s. This is 'high forestry' to produce 'timber' rather than the various forms of 'coppice' and 'coppice-with-standards' that applied traditional techniques to produce both 'wood' and 'timber'. As forestry developed, the approach was exported around the globe by Western European imperial powers. However, this was not one-way traffic since in return the imperial foresters brought back new species of trees to populate their high forest plantations. From the period around the late 1700s to the late 1900s, vast areas of Britain's heaths, moors, bogs, royal forests, and woods, were planted or converted to high forest with exotic species. Many areas were to be dominated by alien conifers, but others by broad-leaves such as sycamore, beech, and imported European oaks. The results today are large areas transformed in terms of landscapes, soils, and ecology, and many of the exotic trees now freely seeding themselves across the wider landscape. Of course, as these often monocultures of alien trees have grown, one by one diseases and pests that thrive in the dense, single-species colonies have followed them.

With radically altered soils and light conditions, the associated ecologies have developed character and distinction, with many native birds, mammals and invertebrates now particularly associated with these exotic habitats. Some flowering plants and numerous fungi have strong associations with long-established conifer plantations.

Not all forestry and tree planting with exotic species was commercially driven, and major landowners, both public and private, undertook much establishment for reasons of 'landscape improvement' and beautification.

Wetland drainage: As described for the Yorkshire region and for the whole of the east of England lowlands (Rotherham 2010, 2013a), widespread land drainage since Roman times, but especially since the medieval, has transformed vast areas. Entire landscapes, particularly of lowland wet woodland, of riverine forest, of lowland fen, marsh and wet heath and bog, have been entirely removed. In their place, we have the modern farming landscapes and especially intensive arable prairies. Whilst some vestiges of the native fauna and flora may hold out in small areas, and especially in nature reserves, this transformation is mostly complete with over huge areas landscapes dominated by alien, hybrid or intensively bred crops. Whilst roadside verges for example, in some areas hold remnants of native communities from the medieval landscape, many are populated by invasive aliens, by highly competitive natives, and by opportunistic species such as roadside halophytes spreading from native saltmarshes inland along linear corridors because of road-salt pollution. Even the linear waterways of field drains and ditches or the larger canals, will have a mix of native and exotic waterweeds, and the alien brown rat will rub shoulders with native water voles. Within the arable fields, 'weed' species both native and long-term aliens formerly did very well. However, adversely affected by modern farming methods and especially herbicides, the diversity of native and archaeophyte arable weeds has collapsed. Today's arable fields do still support floras of weed flowers and associated invertebrates and birds, but again a recombinant mix of natives and aliens.

Urban development: The development of urban centres has happened around the world for thousands of years. However, it was only in the early twenty-first century that humans became predominantly 'urban' with over half the global population now in towns and cities. Where urban settlement occurs, the landscape and its ecologies are transformed. Furthermore, towns and cities act as growth-poles for recombinant ecology as species and people intertwine in a melting pot of urban sprawl and human impacts. Urban development draws into itself people and resources such as animals, plants, and the products from both. People establish gardens and parks with exotic species of plants but often of animal too. The impacts of towns and cities rebound back into the wider environment as pollution as a naturalising fauna and flora.

Additionally, towns and cities change the local environment and particularly the local microclimate, which may trigger potential establishment and invasion by exotic species. This can range from alien ants in buildings, to parakeets in gardens, and Mediterranean figs along watercourses. The urban impact may also create opportunities for invasion by removing established natives and creating vacant or new niches for exotic fauna and flora. The urban footprint extends out beyond the town or city as exploitation of the hinterland triggers transformations of the wider landscape to support the expanding communities. Into this extensive landscape, exotic species such as Himalayan balsam, Japanese knotweed and others, find abundant opportunities.

Trade: A major part of the processes of human colonisation and urbanisation is trade in goods, products, and species – both plants and animals. Not only did people move species with them, but also as Europeans for example, discovered new worlds, they traded and brought back exotic plants and animals from around the planet. Many of these imports perished but some survived, and others did remarkably well to become familiar, often invasive species today. These processes continue today, accelerating as the fashion for gardening has become a national obsession in Britain. New and exotic plants are imported and sold on into gardens via gardening retail outlets and many of these can leap the garden fence. The numbers that actually establish and survive in the 'wild' are limited but with climate change and other environmental perturbations, they are growing. Over time, many of these invaders, such as the medieval imports like snowdrop and sweet cicely for example, tend to settle into the native ecology and to become accepted as 'honorary natives'.

Throughout history, many species of both animals and plants have travelled as unofficial baggage alongside trade to become naturalised into our established ecologies. Some of these, like both black rat and then brown rat, have had profound impacts on both people and native ecology. Specific weed species are associated with for example, cotton cargoes imported from North America, though these are frequently limited in occurrence. Unseen fauna brought in with exotic plants for glasshouse and garden collections, include the micro-arthropod springtails or collembolan. Some species from as far away as New Zealand and Australia can now be found in horticultural collections and spreading into the wider landscape. They are joined by a range of exotic fungi and other taxa, many largely unseen and unnoticed, unless like the New Zealand flatworm they achieve a higher public awareness and profile.

Human migration and colonisation: Finally, one of the major triggers of ecological recombination throughout history has been the movement of people and cultures across the landscape and around the planet. This process continues today but in Britain, probably began in earnest with the arrival of European settlers the Celts. They changed the way that the land looked though farming impacts and technological innovations, and they brought crops, weeds and animals like the brown hare and fallow deer for example, with them. Each wave of colonists since then has marked their arrival with a diversity of new species and ways of using the land that have merged into what was 'natural' as the long-term process of ecological recombination has played out. We can see ecological markers in the landscape of these various waves of cultural integration from the Roman and later Norman introductions of brown hares, of fallow deer, and then the monastic and crusader importations, to the black rat and bubonic plague. Each arrived, established, transformed and merged into the national ecology. The more recent impacts of gardening and of forestry are simply modern manifestations of this established process.

References

Carver SJ (2014) Making real space for nature: a continuum approach to UK conservation. ECOS 5(3/4):4–14

Davis MA, Thompson K, Grime JP (2001) Charles S. Elton and the dissociation of invasion ecology from the rest of ecology. Divers Distrib 7:97–102

Gilbert OL (1989) The ecology of urban habitats. Chapman and Hall, London

Gilbert OL (1992a) The flowering of the cities….The natural flora of 'urban commons'. English Nature, Peterborough

Johnson S (ed) (2010) Bioinvaders. White Horse Press, Cambridge

Monbiot G (2013) Feral: searching for enchantment on the frontiers of rewilding. Allen Lane, London

Prins HHT, Gordon IJ (2014) Invasion biology and ecological theory. Insights from a continent in transformation. Cambridge University Press, Cambridge

Rackham O (1980) Ancient Woodland: its history, vegetation and uses in England. Edward Arnold, London

Rackham O (1986) The history of the countryside. Dent, London

Rotherham ID (2001) Himalayan Balsam—the human touch. In: Bradley P (ed) Exotic invasive species—should we be concerned? Proceedings of the 11th conference of the Institute of Ecology and Environmental Management, Birmingham, April 2000. IEEM, Winchester, pp 41–50

Rotherham ID (2005a) Invasive plants—ecology, history and perception. J Pract Ecol Conserv Spec Ser 4:52–62

Rotherham ID (2005b) Alien plants and the human touch. J Pract Ecol Conserv Spec Ser 4:63–76

Rotherham ID (2007) The implications of perceptions and cultural knowledge loss for the management of wooded landscapes: a UK case-study. For Ecol Manag 249:100–115

Rotherham ID (2008) Lessons from the past—a case study of how upland land-use has influenced the environmental resource. Aspects Appl Biol 85:85–91

Rotherham ID (2009a) Exotic and alien species in a changing world. ECOS 30(2):42–49

Rotherham ID (2009b) The importance of cultural severance in landscape ecology research. In: Dupont A, Jacobs H (eds) Landscape ecology research trends. Nova Science Publishers Inc., USA

Rotherham ID (2010) Yorkshire's forgotten Fenlands. Pen & Sword, Barnsley

Rotherham ID (2011) History and perception in animal and plant invasions—the case of acclimatization and wild gardeners. In: Rotherham ID, Lambert RA (eds) Invasive and introduced plants and animals: human perceptions, attitudes and approaches to management. EARTHSCAN, London, pp 233–247

Rotherham ID (2013a) The Lost Fens: England's greatest ecological disaster. The History Press, Stroud

Rotherham ID (ed) (2013b) Cultural severance and the environment: the ending of traditional and customary practice on commons and landscapes managed in common. Springer, Dordrecht

Rotherham ID (2014a) Eco-history: an Introduction to biodiversity and conservation. The White Horse Press, Cambridge

Rotherham ID (2014b) The call of the wild. Perceptions, history people & ecology in the emerging paradigms of wilding. ECOS 35(1):35–43

Rotherham ID (2016) Eco-fusion of alien and native as a new conceptual framework for historical ecology. In: Vaz E, de Melo CJ, Pinto L (eds) Environmental history in the making, vol 1. Springer, Dordrecht (in press)

Rotherham ID, Lambert RA (eds) (2011a) Invasive and introduced plants and animals: human perceptions, attitudes and approaches to management. Earthscan, London

Rotherham ID, Lambert RA (2011b) Balancing species history, human culture and scientific insight: introduction and overview. In: Rotherham ID, Lambert RA (eds) Invasive and introduced plants and animals: human perceptions, attitudes and approaches to management. Earthscan, London, pp 3–18

Stace CA, Preston CD, Pearman DA (2015) Hybrid Flora of the British Isles. Botanical Society of Britain & Ireland, Bristol

Taylor P (2005) Beyond conservation. A wildland strategy. Earthscan, London

Vera F (2000) Grazing ecology and forest history. CABI Publishing, Oxon, UK

Photograph: Massive flower-heads of giant hogweed are another Victorian garden import, this time from the Caucasus

Chapter 3
The Impacts of Urbanisation

Urbanisation and Recombinant Ecology

Urban areas and urbanisation have long been a focus of those interested in exotic species. From Richard Fitter in 1945, to Gilbert (1989), and more recently Goode (2014), a relatively small group of ecologists has focused on the 'urban', on nature in cities and towns as a counterbalance to work on the rural countryside. Now in the twenty-first century, a central driver of ecological hybridisation is the transformation of human society to a largely urban community, a process that has happened on different scales and over varying periods around the globe. Now for the first time, most humans are urban dwellers and their ranks swell daily. Furthermore, not only do urban landscapes engulf and subsume the rural countryside, but also the loss of people from those areas can trigger massive changes through 'cultural severance' (see Rotherham 2009, 2015a, for example). There is therefore a natural connection between the studies of urbanisation and urban ecologies, and those of novel ecologies and recombinance. Above all, it is vital to recognise that the landscapes and their ecologies, particularly in places like Britain, are eco-cultural and not natural.

In this chapter, the consequences of human urban influences are considered in relation to ideas of plant strategies and related theories of adaptation and evolution. In the urban areas, as discussed in Chap. 4, the footprints of gardeners and horticulturalists have wrought huge changes to present exotic and native ecologies with new opportunities and remarkable challenges. In urban areas and associated with modern urban living, we pollute the land, the air and the waters (see Freedman 1995 for example) thus changing the basic canvas on which ecological processes are acted out. Finally, the dramatic consequences of human cultural severance from nature and the natural world, and the implications for ecology and biodiversity are explained.

© The Author(s) 2017
I.D. Rotherham, *Recombinant Ecology - A Hybrid Future?*,
SpringerBriefs in Ecology, DOI 10.1007/978-3-319-49797-6_3

Urban Centres and Novel Ecologies

Urban areas are characterised by radically modified environments and often-extreme conditions. These may include built structures, artificial surfaces that make existence difficult or impossible for some species but provide opportunities for others. Urbanisation may drive many species to extinction or at least local displacement, but this may leave vacant niches to be filled and nature abhors a vacuum. Furthermore, the processes of disruption and disturbance are radically transformed with patterns of regular and often predictable micro-disturbance replaced by massive and unpredictable macro-disturbance.

Soils are often removed, covered or extremely modified by pollution, rubble and other debris, and loss of water; and if they remain, are frequently eutrophic. In post-industrial sites, there are often extremes of pH from highly acidic to very alkaline, and such sites may have water conditions ranging from severely desiccated to waterlogged. Such conditions provide opportunities for stress tolerant species pushed out of the wider environment by macro-disturbance and eutrophication, and for novel communities of these and of opportunist exotic species. Most of these newly emergent communities are only poorly known and the British National Vegetation Classification or NVC (see for example, Rodwell 2000) does not address urban or post-industrial communities and makes no mention of recombinants. However, when they have been described, (for example, Rotherham et al. (2012) and Rotherham (2012), for post-coaling sites), they have been especially interesting and include rare native species.

Indeed, with so-called post-industrial brownfield environments, there are unique juxtapositions of site histories and both inherited and colonising species that generate communities, which are potentially sustainable, and of high conservation value (Rotherham 1999, 2012). Nevertheless, with the exception in recent years of sites conserved for specially protected species such as great crested newt, and occasionally for rare invertebrates, these areas are mostly overlooked. Yet mixing native and exotic in varied and extreme environmental conditions, some post-industrial sites can become remarkably species-rich and hold regionally significant populations of Red Data Book species (Rotherham 1999). The extreme conditions of post-industrial sites may slow or even halt ecological successional processes and may benefit stress tolerant species and stress tolerant ruderals. In this respect, they differ from most post-housing 'urban commons' which in many cases are highly disturbed, potentially eutrophic, and so proceed quickly to tall herb, scrub, and secondary woodland.

Ecological Strategies and Urban Recombinants

With reference to theories of ecological strategies (e.g. Grime et al. 2007), and ecological successional processes, it is possible to both understand these events and to predict the potential outcomes. Taking the basic model of the three strategic positions of stress-tolerators, ruderals, and competitors, we can better understand the successional and community positions of the individual species in these novel systems. In polluted, post-industrial sites, there are frequently mixes of native stress tolerators and ruderals together with native competitors and ruderals. The communities depend on both the site history and the post-industrial conditions remaining after abandonment, and the nature of the surrounding landscape as a source of seeds and other plant propagules. Beyond this, the communities that develop are influenced by disturbance factors, microclimate, importantly hydrology, and by communities from the pre-industrial landscape absorbed into the site but surviving industrial development. In areas of otherwise intensive agricultural or urban land-use, brownfield sites may hold the last remnants of pre-industrial ecology, but this is seldom recognised.

Along continuums of environmental conditions such as climate, altitude, or proximity to maritime influences, the species colonising into a post-industrial site, and hence the resulting recombinant communities, may vary to reflect those easily available from the local environment. Some species are able to arrive from longer distances and others are essentially dependent on local sources. Similarly, the ability of species to survive through a period of major disruption such as industrialisation may also vary with local conditions. Obvious examples included species primarily associated with woodland environments surviving through non-woodland phases in the Atlantic-influenced western zones of Britain, and generally in coastal areas. It is these colonisers and the survivors that provide the mix for the novel, recombinant communities.

Along with the disruption and pollution of individual sites, industry, urban centres, intensive agriculture, and other human activities, cause widespread pollution of air and water. Both these impacts on air and water, 'the body odour of human urban activities' (Rotherham 2014), are major drivers of eco-fusion leading to emerging recombinant ecologies.

Polluting the Air

Air pollution is an obvious, widespread consequence of industry and urban life and the impacts have been widely documented (e.g. Carson 1965; Mellanby 1967). In past times the air pollution in many British cities, was devastating to people and to ecology (Bornkamn et al. 1982). Often acute but at other times less obvious but with insidious ecological effects, from the removal of lichen floras to the acidification of woodland and grassland soils, air pollution has transformed much of our ecology. However, even in the wider countryside the effects of widespread nitrogen from atmospheric sources is transforming ecology and shifting the balance towards competitive, often catholic species. This is a major example of the eco-cultural nature of ecology and the resultant transformations of eco-fusion to recombinant systems. Some effects are obvious and instantly recognisable; others are less visible but significant nonetheless.

Although some recovery is now taking place, the impact of two centuries of air pollution annihilated populations of pollution-sensitive species in towns and cities and in a wide footprint around them. In Great Britain for example, lichens and plants such as ferns were removed from whole areas of the landscape, both urban and rural as pollution spread widely across the areas down-wind of the sources. From domestic and industrial coal fires, acid rain combined with soot and grime to make conditions intolerable for many plants and animals; ecology and associated biodiversity of many areas irreparably changed. This removed arboreal lichens, and killed sensitive trees, and other species. However, more long-term perhaps, and mostly overlooked, they transformed soils with decreased pH and loss of nutrients such as bases. In cities like Sheffield (Bownes et al. 1991), this has undoubtedly transformed entire ecosystems and caused the massive loss of sensitive plants and animals. Base-loving plants recorded in the coal measures soils of the city in the 1700s can no longer be found except on the Carboniferous and Magnesian limestone areas ten or fifteen miles away. These species distribution are generally assumed to reflect underlying geological conditions with a separation of acid-loving and base-demanding flora by soil type as a reflection of the rocks. In fact, history tells of a landscape and its soils transformed by human activities, and an eco-cultural ecology.

Other effects seen widely across former industrial cities are the deposition in watercourses and on land of large amounts of air-borne, long-residence pollutants such as lead. The implications of such widespread and long-lived pollution are unknown, though so long as they are biological inactive, they are assumed to be benign. Some of the other post-industrial legacies of more active and biologically available toxins, especially heavy metals, in river sediments such as along the River Rother in both Yorkshire and Derbyshire, remain elusive (Rotherham 2008).

The effects of pollution on biodiversity are not simply to reduce or even remove particular species, but sometimes to set in train a longer-term displacement of some species by others. Therefore, the contamination of soil and water by nitrogen-rich, inorganic fertilizers for example, produces conditions most suitable for aggressive and faster-growing competitor species. These displace the stress-tolerators, which require lower levels of available nutrients. One of the most elegant examples of this process is the impact of air pollution on lichens in urban areas (Richardson 1992). Drastic air pollution in English industrial cities progressively removed the pollution sensitive species such as *Usnea* and *Ramalina* and facilitated the appearance and ultimately the dominance of species such as the '*pollution lichen*' *Lecanora conizaeoides*, a species unknown until catastrophic human impact. Indeed, assessments of the selective impacts of air pollution can be applied to use lichens as environmental indicators or '*bio-indicators*' (Richardson 1992; Rose 1976). If air is

badly polluted, particularly with sulphur dioxide given off as gas by burning sulphurous mineral coal, then there may be zones with no lichens at all. At the other
extreme, if the air is clean, then a range of shrubby, hairy and leafy lichens may be
abundant and the diversity of lichens increases dramatically. Some lichens are able
to tolerate relatively high pollution levels, and may be found commonly in on urban
pavements, walls and tree bark. Evaluation of air pollution levels reflects available
substrate too, including the trees and bark available. The acidic bark of trees such as
pedunculate oak (*Quercus robur*) are rather poor for most lichens whereas those
with more alkaline bark such as ash (*Fraxinus excelsior*) or willow (*Salix* sp.) are
far more suitable.

The lichens which are most sensitive to air pollution are shrubby and leafy since
their '*branches*' protrude out into the pollution-laden atmosphere. Those that are
most tolerant tend to be the somewhat structurally reduced forms or crustose
lichens. Following heavy industrial pollution in the late 1700s and 1800s, the more
sensitive shrubby and leafy lichens (*Ramalina, Usnea* and *Lobaria* species), had
limited geographical ranges largely removed from anywhere near to, or down-wind
of, an air pollution source. By the 1970s, they were confined to regions of Britain
with unpolluted air and today are especially abundant in northern and western
Scotland, west Wales, Devon and Cornwall. However, since the decrease in
atmospheric sulphur during the 1980s and 1990s, many of these species have
expanded back into their original ranges (Richardson 1992), an eco-culturally
driven pulsing of the ecology. The ecological pendulum has swung back but into a
slightly different orbit since nitrogen pollution has shifted the returning communities towards more eutrophic-tolerant species.

In terms of biodiversity, human impact has had clear effects on lichen communities around cities. The results of air pollution and modifications to substrate
availability have generated lichen zone patterns visible around towns or cities, and
around individual industrial complexes. The driving force in these ecological zones
is the mean level of sulphur dioxide in the atmosphere and in rain and indices of
pollution levels have been produced (e.g. Hawksworth and Rose 1970). The latter
has a scale of 1 (poorest air) to 10 (purest air) and is a good general '*index*' of
ambient air quality.

However, an interesting observation of human eco-cultural impact on biodiversity is that whilst the Hawksworth and Rose zonation applies well in situations
with rising sulphurous air pollution, as levels fall, the reverse does not apply. With
sulphur dioxide decreasing with for example, air pollution controls, a shift from
coal burning in domestic fires, or with the closure of heavy industrial factories, the
lichens do not re-colonize in the same sequence in which they were lost. The more
random effects of propagule dispersal and substrate suitability probably have
greater influence on the subsequent distribution patterns.

In the early to mid-twentieth century in most British towns and cities, air pollution was greater than it is today. Sulphur dioxide pollution was worst in the inner
city and declined out to the suburbs. In this situation, a scale for lichen pollution
zones would highlight *Zone 1* as the inner city, with progressive improvement to the
cleaner air at the transition to the countryside beyond the town. However, since the

1970s, sulphur dioxide concentrations have been falling in both the inner and outer city zones with far less difference between them. This reduction in sulphur dioxide from the 1970s to the present day has triggered the re-colonization by a number of lichen species. Interestingly too, it has meant a drastic reduction in the pollution indicator *Lecanora conizaeoides* which is now once again becoming quite rare and is generally confined to very acidic tree barks such as those of Scots pine. In the wider landscape, the effects can be confounded by air pollution from traffic and from major power stations beyond the urban area. Human influences are seen to be driving eco-fusion and recombination even at the micro-level of the flora and fauna on tree bark and on other substrates.

In essence, we see the widespread and almost total destruction of these communities and their associates such as invertebrates followed by recovery but to a new series of assemblages. The removal of typical leafy lichens with pale grey, speckled colouration and replacement by dark smoke-stained surfaces and the dominance of 'pollution lichen' *Lecanora*, triggered a well-documented incidence of co-evolution. The pale, speckled moth *Biston betularia*, the peppered moth, quickly 'evolved' a dominant dark form with cryptic colouration to disguise its presence in smoky, sooty, urban environments. As pollution levels dropped and dark surfaces were once again replaced by pale ones, the trend reversed.

Polluting the Waters and the Land

Air has been described as the breath of life, in which case water is the blood of life and the rivers its arteries. In this context, human influences on rivers in Britain, and hence on all the associated ecosystems and species, have been devastating. Transformation, displacement, eradication, and now eco-fusion have generated the highly modified systems and ecologies that we see today. Rivers and other waterbodies have been altered radically from what was here even two centuries ago. Physically, chemically, and biologically they have all been transformed. Over the long period of industrialisation, many species were inexorably squeezed out. Now, in the twenty-first century a recovery is taking place as habitats are patched up and chemical pollution levels drop. This is creating opportunities for a remarkable resurgence and renaissance in for example, urban riverine ecology but as an eco-culturally driven phenomenon. In urban centres with grossly modified physical structures and conditions, native and exotic species of animals and plants vie with each other to establish and succeed in this recombinant environment. In Sheffield's River Don for example, where in the 1970s there were no fish apart from some bizarrely mutant or at least malformed sticklebacks, there are now at least twelve species including pollution-sensitive ones like brown trout. Native river plants have been reintroduced along with appropriate sports fish. However, anglers also introduce to rivers and other wetlands, a host of non-native species that they want to see back in the now clean river, and many of these establish and naturalise. Predatory birds like kingfisher, grey heron, and cormorant do not discriminate between native and exotic as they scour the river for prey. Similarly, the native otters and exotic American mink hunt both native and alien fish, and seek cover under largely invasive tall herbs like Japanese knotweed and Himalayan balsam.

The precise impacts of river pollution depend on many factors including the type of pollutant, the nature of the river, and capacity to absorb or recover from pollution incidents (Mellanby 1967). Furthermore, the impacts of pollution on biodiversity vary with season and hence vulnerability of particular organisms. The impacts vary between regular repeated pollution releases and one-off incidents. The range of pollutants and their effects is infinitely wide and varies from relatively benign materials such as sewage, which simply use up available oxygen in their degradation, to highly toxic chemicals such as mercury, arsenic and dioxin, with potentially long-term impacts. This means the effects on aquatic ecosystems are similarly varied. However, just as with the air pollution zonation found with lichens and sulphur dioxide, there are specific and predictable patterns of biodiversity in polluted waters. Through detailed understanding of these patterns, quite sophisticated assessments can be made of the levels of water pollution occurring at a site over time. These effects are seen in microorganisms such as bacteria, algae, blue-green algae, and fungi, and especially in aquatic arthropods (such as crustaceans and insects with aquatic larvae), and in fish. Just as with air pollution in cities, the pollution can squeeze species out and then if conditions recover, a process of re-colonisation may occur (Mellanby 1967).

As was the case with the lichens and air pollution, the grossly degraded waters gained new and distinctive assemblages of flora and fauna that were distinctive from what went before. In particular, in river zones with high levels of organic effluents and hence low dissolved oxygen concentrations, the typical fauna and flora were displaced and a new ecological mix was assembled. This was dominated by a slimy growth termed 'sewage fungus', which was in fact a mass of filamentous bacteria of the genus *Sphaerotilus*; and associated with this was a very distinctive but limited fauna of pollution tolerant invertebrates. In waters with high levels of nitrogen pollution, the dominant flora was made up of filamentous and blue-green algae.

In the wider countryside, as first described in the 1960s (e.g. Carson 1965), agricultural pesticides have had and still have widespread and pernicious impacts on ecological systems. However, the disruption of rivers and other waterbodies downstream of towns and cities can be equally dramatic and from Sheffield in the 1970s, the sewage-related ammonium pollution into the River Don was detectable at the mouth of the great Humber Estuary downstream nearly 100 miles away. These arteries of the wider ecosystem hold radically altered recombinant ecologies responding to long-term pulses of human disruption through the processes of eco-fusion of natives and exotics. Removal and extinction followed by chemical recovery trigger opportunities for re-colonisation and the generation of novel ecological communities. Some of these systems may have strong similarities to former ecologies, but they will have mixes of different species, altered distributions, and even changed keystone components. At the mouth of the estuary for example, the sediment patterns and the consequent food webs, will be influenced strongly by the hybrid cord-grass *Spartina* x *townsendii*, or the associated species complex, and not by a native.

Human impact on rivers and streams is not simple. Physical disruption through canalisation, culverting, diversions, and increases in sediment loads and deposition patterns, can all affect the ecology. In some cases, these impacts may be at least as significant as the levels of pollutants. However, the effects can be dramatic as well as unexpected. In industrial cities in northern England, such as Bradford, Sheffield, Leeds, and Manchester, the combination of chemical, physical and biological pollution plus the structural manipulation and modification of many urbanised watercourses led to the almost entire removal of all biodiversity from entire stretches of river and stream. In some cases, like the Derbyshire River Rother, slugs of long-term industrial pollution can be tracked as they move slowly downstream (Rotherham 2008). Indeed, through intensive industrialisation these systems and the banks and floodplains immediately adjacent were generally degraded and despoiled and in some cases biologically dead (Bownes et al. 1991; Gilbert 1989). Industrial degradation, contamination and despoliation transformed entire landscapes, removing much of the former ecology. However, in the post-industrial era, the chemical recovery has created opportunities for new recombinant ecosystems (Rotherham 1999; Rotherham et al. 2003, 2012). Even the construction of walls, in urban and other environments, creates analogues for natural features and the opportunities for new and ecologically fused communities (Gilbert 1992b).

As the pollution levels dropped, plants and animals re-colonised but with a further human, eco-cultural twist to the biodiversity mix. Exotic plants sometimes escaping from the former Victorian '*wild*' gardens, to which plant collectors had introduced them, slipped out almost unnoticed into the newly feral landscape (Rotherham 2005a, b). In some cases, they were assisted by deliberate introductions by enthusiasts for exotic plants. In twentieth-century urban watercourses, once almost devoid of life just as urban areas became lichen deserts, as pollution levels dropped, developed a new recombinant flora. This new ecology is dominated by alien trees such as sycamore (*Acer pseudoplatanus*) with aggressive perennial herbs like Japanese knotweed (*Reynoutria japonica*), robust biennials such as giant hogweed (*Heracleum mantegazzianum*) and vigorous annuals particularly Himalayan balsam (*Impatiens glandulifera*).

As Oliver Gilbert demonstrated in the 1980s (Gilbert 1989, 1992a, c), these dense stands of exotic trees and herbs were forming new and distinctive urban plant communities, though perhaps with more human influence than Gilbert imagined. He also demonstrated the remarkable re-colonisation of native plants typical of ancient woodlands, washing downstream from wooded sites in the headwaters to establish '*ancient woodland*' floras beneath pseudo-canopies of dense knotweed and balsam (Gilbert 1989, 1992c). Thus, we witness the establishment of novel ecological and recombinant systems triggered by unique sequences of human-induced environmental changes; a further demonstration of the intimate relationships between human influences and contemporary biodiversity.

Urban rivers did not just suffer chemical pollution, alteration of built structures (riverbanks etc.), and massive inputs of human sewage effluent, but some became warmer too. In fact, industrial urban rivers such as the Don in Sheffield used to cool the great factories such as steelworks had their ambient temperatures raised to a more-or-less constant 20–23 °C, winter and summer. This situation continued until the 1970s, when many factories closed and others became more self-contained and rigorously controlled. However, combined thermal pollution and widespread sewage until the late 1970s, brought together seeds of wild Mediterranean figs (*Ficus carica*) with the conditions needed for germination. In a richly eco-cultural history, the fig seeds pass through the human gut unaltered and ready to germinate if given constant warm, moist conditions. With abundant raw sewage spilling from Victorian sewers, they established along many urban rivers, though most famously in Sheffield. These figs now sucker and grow and in some areas have formed unique '*urban fig forests*' along our most urban watercourses. In Sheffield, these trees are especially protected as iconic symbols of the Industrial Revolution and a particularly exotic contribution to regional biodiversity (Bownes et al. 1991; Gilbert 1989).

References

Bornkamm R, Lee JA, Seaward MRD (eds) (1982) Urban ecology. Blackwell Scientific Publications, Oxford

Bownes JS, Riley TH, Rotherham ID, Vincent SM (1991) Sheffield nature conservation strategy. Sheffield City Council, Sheffield

Carson R (1965) Silent spring. Penguin Books Ltd., Harmondsworth, Middlesex

Freedman B (1995) Environmental ecology—the effects of pollution, disturbance and other stresses, 2nd edn. Academic Press, San Diego

Gilbert OL (1989) The ecology of urban habitats. Chapman and Hall, London

Gilbert OL (1992a) The flowering of the cities…. The natural flora of 'urban commons'. English Nature, Peterborough

Gilbert OL (1992b) Rooted in stone. The natural flora of urban walls. English Nature, Peterborough

Gilbert OL (1992c) The ecology of an urban river. British Wildlife 3:129–136

Goode D (2014) Nature in towns and cities. Collins New Naturalist, London

Grime JP, Hodgson JG, Hunt R (2007) Comparative plant ecology. A functional approach to common British species, 2nd edn. Castlepoint Press, Dalbeattie

Hawksworth DL, Rose F (1970) Qualitative scale for estimating sulphur dioxide air pollution in England and Wales using epiphytic lichens. Nature 227:145–148

Mellanby K (1967) Pesticides and pollution. Collins, London

Richardson DHS (1992) Pollution monitoring with lichens. Naturalists' handbooks no. 19, Richmond Publishing Co. Ltd., Slough, UK

Rodwell JS (ed) (2000) British plant communities: volume 5, maritime communities and vegetation of open habitats: maritime communities and vegetation of open habitats. Cambridge University Press, Cambridge

Rose F (1976) Lichenological indicators of age and environmental continuity in woodlands. In: Brown DH, Hawksworth DL, Bailey RH (eds) Lichenology: progress and problems. Academic Press, London

Rotherham ID (1999) Urban environmental history: the importance of relict communities in urban biodiversity conservation. Pract Ecol Conserv 3(1):3–22

Rotherham ID (2005a) Invasive plants—ecology, history and perception. J Pract Ecol Conserv Spec Ser 4:52–62

Rotherham ID (2005b) Alien plants and the human touch. J Pract Ecol Conserv Spec Ser 4:63–76

Rotherham ID (2008) Landscape, water and history. Pract Ecol Conserv 7:138–152

Rotherham ID (2009) The importance of cultural severance in landscape ecology research. In: Dupont A, Jacobs H (eds) (2009) Landscape ecology research trends. Nova Science Publishers Inc., USA

Rotherham ID (2012) Post coal-mining landscapes; water, heaths, and commons as a resource for wildlife, people and heritage. Landscape archaeology and ecology special series (2), between a rock and a hard place, pp 49–58

Rotherham ID (2014) Eco-history: an introduction to biodiversity and conservation. The White Horse Press, Cambridge

Rotherham ID (2015) Bio-cultural heritage & biodiversity—emerging paradigms in conservation and planning. Biodivers Conserv 24:3405–3429

Rotherham ID, Spode F, Fraser D (2003) Post–coalmining landscapes: an under-appreciated resource for wildlife, people and heritage. In: Moore HM, Fox HR, Elliot S (eds) Land reclamation: extending the boundaries. Published: A.A. Balkema Publishers, Lisse, pp 93–99

Rotherham ID, Lunn J, Spode F (2012) Wildlife and coal—the nature conservation value of post-mining sites in South Yorkshire. In: Rotherham ID, Handley C (eds) Dynamic landscape restoration. Landscape archaeology and ecology special series. Papers from the landscape conservation forum, vol 1, pp 30–64

Photograph: Ring-necked parakeet—a relatively recent, twentieth century import and spreading rapidly and in large numbers (photograph by Andy Waple)

Chapter 4
The Impacts of Globalisation and Cultural Severance

The Impacts of Cultural Severance

One major driver of change over time is the process that I have defined as 'cultural severance' (Rotherham 2009, 2013). This is not necessarily an urban process, but occurs extensively in the wider rural environment and is associated with rural de-population and migration to the towns and cities. In this context, cultural severance is strongly associated with the socio-economic drivers of urbanisation and of globalisation and in the wider countryside may trigger landscape abandonment on the one hand and agro-industry on the other. In essence, this involves the breakdown of utilitarian subsistence uses and dependence on local natural resources and the distancing of people from 'nature'. These processes have occurred over long time-scales but in Britain, they have accelerated since the 1700s and 1800s. The impacts on ecology have been dramatic, in particular with trends of eutrophication, macro-disturbance, habitat and species displacement, and the loss of traditional or customary management of resources. These changes have then triggered major shifts towards exotic species in recombinant ecosystems.

Along with urbanisation is the major influence of globalisation of people, society, economies and of nature. A response to these tensions and changes is the emergence of a new recombinant ecology, often but not always, evolving in urban centres. Interestingly too as globally we become more urban, it is this ecology with which most people now interact. People, species, and resources move rapidly around the planet and the impacts are seen as new pests and diseases wipe out native trees, and invasive alien plants and animals displace long-established natives. Human cultural influences take effect too and ideas and fashions shape desires for landscapes and for gardened nature. Globalisation implies and indeed demands, mobility; and this now forges contemporary ecology through eco-fusion processes.

Intertwined with urbanisation, globalisation, and rural depopulation, cultural severance is manifested through the widespread ending of traditional and customary

© The Author(s) 2017
I.D. Rotherham, *Recombinant Ecology - A Hybrid Future?*,
SpringerBriefs in Ecology, DOI 10.1007/978-3-319-49797-6_4

management of the land and its resources. The consequences for ecology include the mixing of species, both alien and native to a region, and the removal of human interventions, which attenuate ecological successional processes. Combined with eutrophication and macro-disturbance to ecosystems, the result is a rapid shift to catholic, competitive and ruderal species and opportunities for invasive exotics, all leading to recombinant systems.

Global Empires, Acclimatisation and Wild Gardening

Until the 1940s and the aftermath of the Second World War, Britons travelled the world seeking out and collecting new species to take home; a remarkable example of culturally-driven eco-fusion. Indeed, they went a step further and implemented deliberate programmes of release into the countryside to 'improve it'. This was intended to bring about economic gain on the one hand and to beautify the land-scape on the other. The ideas spread and a movement emerged and was formalised. These fashions then spread around the world, firstly with European Acclimatisation Societies, aiming to introduce and test new economic crops and to investigate the potential for food. In Britain and the British colonies, the Acclimatisation Societies sought to introduce animals and birds to new territories to better their economies, gastronomies, and landscapes (Lever 1977; Rotherham 2011). These Victorian Acclimatisation Societies had massive impacts around the planet in places such as New Zealand, Australia and the West Indies for example. Indeed, in many places these introduced species transformed the landscape and eliminated many natives.

Whilst it is sometimes suggested that the British Acclimatisation Societies had only a limited impact over their short lifespan, I suggest that this is not so. The effect on British ecology was greatest through the changing attitudes and the ideas of 'improvement' to native faunas, and led to a diversity of successful species introductions over a period of nearly a century. Many of these are still with us and some are important in contemporary ecosystems:

Sika deer (*Cervus nippon*)
Muntjac deer (*Muntiacus muntjak*)
Red-necked wallaby (*Macropus rufogriseus*)
Grey squirrel (*Sciurus carolinensis*)
Little owl (*Athene noctua*)
Canada goose (*Branta canadensis*)
Egyptian goose (*Alopochen aegyptiacus*)
Mandarin duck (*Aix galericulata*)
Ruddy duck (*Oxyura jamaicensis*)

Some of these were deliberately 'naturalised' into the 'wild', but others, kept freely in collections such as grand, landscape parks, simply leapt the perimeter fence to freedom. Numerous other species either escaped or were released but failed to establish. These include Himalayan porcupine (*Hystrix hodgsonii*), American bison (*Bison bison*), and other exotics.

Apparently neglected by authors such as Stace and Crawley (2015), the 'Wild Garden' movement as promoted by William Robinson (1870), had a massive and obvious impact on alien invasive plants and subsequently on British recombinant vegetation. Whilst not a formal movement like the Acclimatisation Societies, this changed approaches to growing exotic flora transformed gardens. From these strongholds established across wide areas, both urban and rural, potential invaders escaped to the wild or were introduced deliberately or accidentally via garden refuse dumped into hedgerows, woods etc. The British became (and indeed remain), obsessed by the gardening of exotic species, and as discussed below, many of these became invaders. From lakes, ponds, and aquaria a diversity of invasive aquatic plants either escaped or else were released into pond, lakes, rivers and canals. Gardeners are often reluctant to kill their excess plants and often prefer to release them into green spaces instead. Furthermore, as discussed in Chap. 3, due to industrialisation and associated pollution and environmental degradation, many plants found their local 'wild' ecosystems already depauperate in species and therefore with abundant vacant niches to occupy (a point noted by Oliver Gilbert). Once the levels of pollution began to drop and habitats became more amenable to occupation, the invaders were well set to thrive.

In Britain in particular, with an imperial domination of the globe and at home a growing penchant for gardening and landscaping, the scene was set for a radical transformation of the countryside and its ecology. Plants such as rhododendron (introduced 1764), Himalayan balsam, Japanese knotweed, and giant hogweed were naturalised, widely across the British countryside Victorian by 'Wild Gardeners'. Foresters and estates managers who rushed to adopt exotic species in their plantations joined gardeners in this transformation of British ecology. In forestry, the European fashion for high forest plantations relied heavily on exotic species. Garden designer and writer William Robinson formalised and popularised the idea of '*naturalising*' exotic plants into landscapes instead of merely planting them for effect (Robinson 1870). Undoubtedly, the use of exotics was already widespread, but Robinson made it mainstream and popular in the 1800s and early 1900s.

Some of these species were soon recognised locally as invaders at a relatively early date, for example Himalayan balsam in Manchester by the mid-1800s (e.g. Grindon 1859), but only as conservation problems much later, by the 1970s and 1980s. However, a whole raft of others such as Montbretia, Buddleia, and Cherry Laurel has now joined them. It seems that these escapees spread across the land and built slowly over perhaps 50–100 years as they colonised suitable environments. However, they then moved into an exponential and explosive expansion at which point botanists and others began to notice and record them. However, by this time, it was of course, almost impossible to control further spread.

More recently, the variegated yellow archangel, which I now believe to be a Victorian cultivar, has been recognised as an active invader. Whenever introduced to woodland or hedgerow it spreads rapidly, covering several hundred square metres of woodland ground flora in only 10–15 years. Since it does not produce viable seed, such spread is wholly vegetative, and its arrival is always by direct human agency. Often colonising into the heartlands of native ancient woodland, this is a dramatic case of recombinant ecology in process. Yet the role of the 'Wild Gardeners' has not been recognised in triggering twentieth-century plant invasions, though with the exception of the massive establishment of exotic forestry conifers, it is at the heart of most terrestrial plant naturalisations in Britain. This was first noted by Rotherham (2001, 2005a, b, 2011), but the reader should seek out William Robinson's hugely influential book 'The Wild Garden' (1870) in order to appreciate the cultural significance of this period. Written by the most popular writer on English gardening in the late 1800s, this landmark volume sparked a revolution in garden taste over the subsequent fifty years or so, and associated with that, an almost unprecedented episode of eco-cultural, ecological recombination.

From the 1700s to the early 1900s, releases of alien plants were firstly into domestic and landscape gardens, and then into forest and woodland estates. It seems that inspired by William Robinson, the Victorian Wild Garden movement was responsible for the introduction and subsequent escape of many plants into the British countryside. Indeed, these include some of the most spectacular invaders such as giant hogweed, and giant knotweed; and these are truly stunning garden plants for the 'wild garden' of the late 1800s. In the mid Victorian period, when William Robinson published his book 'The Wild Garden' in 1870, across the British countryside and in suburban gardens practices he described and advocated were already well established. In particular, these included the embellishment of woodlands, utilising naturalisation of exotics in landscapes, and mixing plants from sub-tropical gardens with emphasis on strong foliage groupings; the impacts remain with us today.

However, with the advocacy of Robinson, this became basis of '*wild gardening*' during the 1870s and 1880s (Elliott 1986). The approach involved large scale, establishment of species capable of spreading themselves in large masses. In this way, plants like *Rhododendron ponticum*, *Heracleum mantegazzianum* (giant hogweed), *Polygonum* sp. (Japanese and giant knotweeds), and Himalayan balsam, were all excellent choices. However, these same wild garden favourites are now the scourge of nature conservation. In situations like Chatsworth Park in Derbyshire or Clumber Park in Nottinghamshire, landscape designers juxtaposed exotic garden plants with forest trees and wild undergrowth.

Today at Chatsworth for example, the rhododendrons, balsam and even giant hogweed remain. William Robinson asserted that the principle of wild gardening was '...naturalizing or making wild innumerable beautiful natives of many regions of the earth in our woods, wild and semi-wild places, rougher parts of pleasure grounds, etc.' Today, an assessment of British exotic flora confirms how many of today's most problematic alien 'weeds' have their origins in deliberate introductions. With globalisation, urbanisation and climate change, the escape of exotic garden plants continues apace.

However, a key factor that is so far overlooked was the extraordinary socio-economic changes that swept through the British countryside from the 1890s to the 1930s. First, there was a change in fashion from great estates and country houses, to more modest urban dwelling. This combined with political forces that led to punitive inheritance tax on property for the very rich and which led to the closure and break-up of many country estates. A system of rural land management with its origins in the medieval landscape was falling apart.

The environmental impacts of these dramatic changes were compounded by the enclosures of common land, the de-population of the countryside and the growth of urban areas. Finally, the effect of the First World War in de-populating rural lands and consigning many great estates to history was the tipping point. With extensive estates and gardens reduced or derelict, a de-populated countryside, and the growth of modern industrial agriculture and forestry, the alien invaders were able to escape into a wider and now hospitable landscape. The teams of gardeners that had once tended the great estates had gone, and there was no longer any barrier to wider colonisation. At first, in fact for many decades, this quiet invasion of the British countryside was largely ignored and the impacts were only being generally recognised in the 1970s. Since that time, though too late for any action to effectively control the spread, many of these plants were identified as the most troublesome invasive weeds in Britain. What happened was in fact a culturally determined eco-fusion of natives, non-natives and archaeophytes to generate new recombinant ecologies across much of the British landscape. The process still goes on.

Another key factor in this successful invasion was the changing of land-use priorities. Significant, in the wider landscape during the period was a change from traditional coppice management of native woods to high forestry. At the same time, commons were enclosed and commoners displaced, with a new, intensive, modern farming landscape imposed over large areas. In woodlands, this new approach was supplemented by introduced shrubs (such as rhododendron, snowberry, *Mahonia*, *Gaultheria*, and other species), and exotic trees from around the world. Many of these were introduced for woodland management for game preservation alongside timber production.

This period of the 1700s to the early 1900s, was perhaps a golden age of plant introductions to the British Isles. There is no going back and since then, there has been a continual release of new exotic species into Britain's landscape from gardens and from modern plantings of exotic trees and shrubs into landscaped greenspaces and along highways. The latter processes, led by the trends and whims of landscape architects, have triggered a new, highly predictable wave of invaders. A wide range of berry-bearing trees and shrubs is rapidly establishing throughout woods and forests, Ratcliffe (1984) noting the beginnings of the spread of shrubs like *Berberis* and *Cotoneaster*. Exotic cultivars of holly and *Sorbus* are now widely established in woodlands, and planted trees such as Norway maple (*Acer plantanoides*) are invading into many landscapes. Processes of eco-fusion and species hybridisation (deliberate and accidental) have created a melting pot of ecological recombinance throughout much of the landscape.

A History of Animal and Plant Importations

The Victorian 'Acclimatisation Societies' and 'Wild Gardeners' (Rotherham 2005a, b, 2011), were manifestations of processes that evolved in varying degrees over centuries. Throughout history, waves of settlers or conquerors of the British Isles had brought new plants and animals with them. The Celts, the Romans, and then Normans and later the Crusaders, for example, imported animals and plants, many of which became keystone species of modern ecologies. Most obvious is the humble rabbit, which is hugely influential in modern landscapes and in many cases is a keystone species. However, fallow deer and brown hare are others and these raise interesting issues of perception and attitude (Rackham 1986).

Alongside deliberate introductions, were many other animals and plants, which escaped domestication to make their own way in the globalised world. Many such species have become intimate components of what is now considered as 'British' ecology. Romans and Normans brought herbs and food-plants from southern Europe and the Mediterranean, as did Crusaders and various monastic dynasties, which controlled large areas of Britain's productive countryside over several centuries. Over this period, many species were absorbed or hybridised into native ecology. Interestingly, most of these species are now tolerated, and many (like the brown hare for example), are celebrated and conserved as 'honorary natives'.

The process continued unabated throughout the later medieval period since by the 1500s, traders and seafarers from Britain and Holland in particular, were charting and colonising the world. From their visits, they brought back exotic plants and sometimes animals; many introductions perishing but not, and accidental imports already included black rat and brown rat, with the added bonus of bubonic plague. In return, later explorers spread dogs, cats, and much more; doing untold damage to previously isolated island ecologies. The cultural homogenisation of ecology was accelerating with the collection and dissemination of exotic species as travellers sought new plants and animals for gardens and menageries. Then, as landscaping, forestry, and gardening erupted in Britain during the 1700s and 1800s, environmental impacts increased. This process continues today sometimes with catastrophic results (Rotherham 2014).

However, it should not be imagined that these changes to 'native' ecology were isolated from other impacts. At about the same time, in the eighteenth and nineteenth centuries, the wider landscape was traumatised by parliamentary enclosures with commonland wrested from the commoners, the peasants and the poor, and converted into intensive food production units. In Scotland, large areas were de-populated by the 'Highland Clearances'. Much of the more natural landscape and its ecology were swept away by this sea of change (Rotherham 2014). Traditional coppice woods were converted to high forest plantations, and industrialising cities began to sprawl across the countryside (Rackham 1980, 1986). Lands, which remained relatively untouched by this were often blended into leisurely landscapes for the pleasure of the landowners and industrialists, and were generally populated by the exotic plants and animals imported from around the world (Rotherham 2014). Associated with the changes was a seminal undercurrent of transformation, from ecology dominated by native 'stress tolerators', often to exotic species of 'ruderal' and 'competitive' plants. This trend was noted for the twentieth century by Davis et al. (2001), but in reality began much earlier as landscapes flexed and changed, and disturbance plus nutrient enrichment came to the fore (Rotherham 2014).

References

Davis MA, Thompson K, Grime JP (2001) Charles S. Elton and the dissociation of invasion ecology from the rest of ecology. Divers Distrib 7:97–102

Elliott B (1986) Victorian gardens. B.T. Batsford, London

Grindon LH (1859) Manchester flora [No publisher or location given]

Lever C (1977) The naturalized animals of Britain and Ireland. Hutchinson & Co (Publishers) Ltd, London

Rackham O (1980) Ancient woodland: its history, vegetation and uses in England. Edward Arnold, London

Rackham O (1986) The history of the countryside. Dent, London

Ratcliffe DA (1984) Post-medieval and recent changes in British vegetation; the culmination of human influence. New Phytol 98:73–100

Robinson W (1870) The wild garden. The Scolar Press, London

Rotherham ID (2001) Himalayan balsam—the human touch. In: Bradley P (ed) Exotic invasive species-should we be concerned? Proceedings of the 11th conference of the institute of ecology and environmental management, Birmingham, April 2000. IEEM, Winchester, pp 41–50

Rotherham ID (2005a) Invasive plants—ecology, history and perception. J Pract Ecol Conserv Spec Ser 4:52–62

Rotherham ID (2005b) Alien plants and the human touch. J Pract Ecol Conserv Spec Ser 4:63–76

Rotherham ID (2009) The importance of cultural severance in landscape ecology research. In: Dupont A, Jacobs H (eds) Landscape ecology research trends. Nova Science Publishers Inc., USA

Rotherham ID (2011) History and perception in animal and plant invasions—the case of acclimatization and wild gardeners. In: Rotherham ID, Lambert RA (eds) Invasive and introduced plants and animals: human perceptions, attitudes and approaches to management. EARTHSCAN, London, pp 233–247

Rotherham ID (2013) The lost fens: England's greatest ecological disaster. The History Press, Stroud

Rotherham ID (2014) Eco-history: an introduction to biodiversity and conservation. The White Horse Press, Cambridge

Stace CA, Crawley MJ (2015) Alien plants. HarperCollins, London

Photograph: Spanish blue bells and the various hybrids are favourite escapes from gardens

Chapter 5
Climate Change and Ecological Hybridisation

It would be inappropriate here to discuss issues of climate and climate change in detail, and no significant contribution to the debate would be made. However, this is such a major influence of past, present, and future ecologies, that a mention and consideration of some pertinent aspects of this are necessary. The accepted fact that we are now in a phase of major climatic flux, no matter how caused, has huge implications for the processes of eco-fusion. Thompson (1994) attempted to predict the impacts of climate change on temperate species, and then Thomas et al. (2004), discussed issues of extinction risk from climate change and associated environmental stresses. These papers and subsequent efforts to model and predict provide a basis to begin to understand possible trends. Walther et al. (2009) for example, consider the specific issues of alien and exotic species in a warmer world. The context for these tensions in global ecology reflected down to regional and national levels, was set by papers such as Rockström et al. (2009) where they attempted tp identify and quantify planetary boundaries that help restrict human activities from causing unacceptable environmental change. They assert that such thresholds should not be transgressed without huge risk for the planet and for humanity. The paper presents a novel approach proposed to define the preconditions for future human development and state that to cross particular biophysical thresholds could have disastrous consequences for the future. Furthermore, they suggest that three of their nine interlinked planetary boundaries have already been overstepped. Whilst they accept that the planet has undergone numerous phases of major environmental change in the past, for the past 10,000 years, the global environment has been relatively and unusually stable. This situation, they argue, is now changing; and I suggest that the consequences of eco-fusion need to be factored into any future ecological predictions.

© The Author(s) 2017
I.D. Rotherham, *Recombinant Ecology - A Hybrid Future?*,
SpringerBriefs in Ecology, DOI 10.1007/978-3-319-49797-6_5

Climate experienced locally as 'weather' affects species distributions and the processes of colonisation. The detail of these interactions varies from species to species but in many cases is due to three main factors:

(1) Winter cold—affecting survival;
(2) Summer warmth—affecting reproduction;
(3) Indirect influences of pests, diseases, and of competition.

These factors have major influences on the outcomes of eco-fusion and the resulting recombinant ecologies.

Many species have climate as a major determinant factor in their biogeographic distributions. Indeed, with access to detailed time-lines of palaeo-ecology and palynology, we can now reconstruct how vegetation and ecology have fluxed over centuries and millennia. Although our view of the world and its ecology is one of stability and predictability, the reality over long periods is anything but. Plants, animals and other organisms respond to short-term weather and to long-term climatic shifts, and this helps shape ecologies and communities.

Climate has always changed and popular ideas of stability are probably reflections of long-evolved needs and individual memories shaped by childhood. Nevertheless, the rate and scale of current climate change is extraordinary, and I argue, is a mix of natural and anthropogenic change. The consequences for a globalising and urbanising ecology are dramatic and challenging. Climate change affects the basic environmental template on which species interactions of competition, succession, colonisation and displacement are acted out. Indeed, it always has done. However, today the impacts of climate and of extreme weather are influencing a radically modified environment and a highly recombinant, globalised ecology.

We begin to see some of the potential impacts of climate-related changes in species and in communities when observing changes in urban ecosystems where heat islands and industrial thermal pollution occur. In the warmth of the city, exotic species that perhaps cannot survive or thrive in the wider environment can do well. The warm, urban River Don in Sheffield for example, displays an ecologically recombinant mix of natives and exotics. A shift in climate may enable invertebrates to survive or allow flowering plants to set seed. Sometimes climate-associated barriers to invasion are removed by the horticultural industry. Invasive water hyacinth for example, is a problem invader on southern waterways in England but has long been restricted in its range by winter cold. Now however, the horticultural retailers offer gardeners 'frost hardy' varieties!

If we consider many of the exotic plants from around the world now grown in British gardens, the controlling factors on their ability to set seed and to spread are often related to winter chill and particularly to frost. As climate warms and frosts become less deep, less frequent, less long lasting, more species will have the potential to seed and to naturalise.

Similarly, many pests and diseases are restricted by harsh winter weather and again may spread more aggressively in a warmer climate. Additional, new invasive diseases and pest species are also likely to occur with the combination of globalisation and climate change.

A further issue is that, as climate change becomes more acute, established native ecologies and species will be increasingly stressed and this will create opportunities for invasives, pests and diseases. It is clear, that with the expected changing scenarios of climate and other environmental conditions, ecological fusion will accelerate and recombinant ecologies will be increasingly apparent. Furthermore, as climate changes and it undoubtedly is doing, established native species may be pushed to the limits of or beyond their environmental envelope, their niche. Then, in the modern fragmented landscape, whilst some species are able to move considerable distances to new islands of suitable habitats, some cannot. Local extinctions may provide vacant niches and hence opportunities for invasive and exotic fauna and flora. In this way, we can expect recombinant communities to spread, and indeed, for the delivery of some ecosystems services, such replacement may be necessary.

Over long periods, fluxes in climatic conditions have driven natural successional processes, ecological fusion, and ecosystem hybridisation, as species across Europe for example, moved southwards and then northwards again, during glacial and post-glacial episodes. These mass movements were accompanied by colonisation, displacement, extinctions and evolution, with communities and ecosystems waxing and waning. Now, with natural changes of climate exacerbated and modified by human-driven processes, the communities and ecosystems are complicated by the impacts of human-induced extinctions on the one hand, and globalisation on the other. The results are increasingly hybrid ecosystems and recombinant ecologies but influenced and driven by the same ecological forces which have always come into play, but moving into new trajectories to reflect environmental change such as climate and opportunity.

Returning to Rockström et al. (2009), Rob Francis suggests (pers. comm.), that how recombinant communities fit into to the assessments of biodiversity thresholds will become increasingly important. What is more, this importance will increase with every passing decade of growing environmental stress and instability, be that climate or other manifestations of change.

Changing environmental conditions clearly help trigger invasions and recombination. Many species of plants and animals have their biogeographical distributions set or curtailed by quite specific bioclimatic limits and if these change so may their distributions. This is nothing new but a core driving process of ecology since life first evolved. The situation now is that the environmental changes are speeded to a degree not seen since one of the 'natural' phases of catastrophe and global mega-extinction. For invasive species and recombination, climate change for example, effectively stirs the pot of diversity and some species mix in whilst others disappear. Climate change may enable the arrival and spread of new species to the region but at the same time it my squeeze the distributions of natives and thus facilitates invasion. These complex processes are worthy of more detailed investigation.

References

Rockström J, Steffen W, Noone K et al (2009) A safe operating space for humanity. Nature 461:472–475

Thomas C et al (2004) Extinction risk from climate change. Nature 427:145–148

Thompson K (1994) Predicting the fate of temperate species in response to human disturbance and global change. In: Boyle TJB, Boyle CEB (eds) NATO advanced research workshop on biodiversity, temperate ecosystems, and global change. Springer, Berlin

Walther G-R, Roques A, Hulme PE et al (2009) Alien species in a warmer world: risks and opportunities. Trends Ecol Evol 24:686–693

Photograph: Wild figs on the urban River Don—an accidental introduction from the Mediterranean

Chapter 6
Future Nature and the Consequences of Recombination

The ideas proposed here present new concepts and paradigms for environmental scientists, ecologists, and nature conservationists. Indeed, in a political and cultural context where it seems that much of the 'environmental debate' has been distilled down to climate change and climate change to anthropogenic carbon dioxide, this approach places new emphasis on the need to understand natural processes better. Furthermore, it is argued that we are witnessing but not recognising dramatic shifts in the ecological template; the long-term consequences of which are significant and substantial. They also present a challenge of recognising and even accepting that some of the changes will happen whether or not we wish this to be the case, and that they are not all bad. History tells us that ecologies have always changed and that humans have frequently been major drivers of the processes.

However, and importantly, it is also explained that species native to an area will also have a place and they retain their conservation importance even in the most urban of environments. The idea that urban environments are somehow different from those beyond the city, and that native species cannot, or even should not, have a place in the recombinant ecosystem, is found wanting. Recombinant ecology includes people both as drivers of change and as actors and participants in the hybridisation processes. Increasingly, it is in the urban environment where most people experience important contact with nature. This urban 'nature' is both native and exotic, reflecting history and projecting future ecologies. We know now that people need, and benefit from, contact with nature. Yet the nature we experience will be a future nature, which is recombinant and hybrid, native and exotic.

Particularly in towns and cities, but also in and around major industrial and post-industrial landscapes, there are new, recombinant ecologies forming. Indeed, some of these are distinctive and with significant associated nature conservation value, yet they are rarely recognised as such. Examples might include the *Buddleia davidii* scrub and woodland of many urban commons sites. This has distinctive and diverse associated herb layers of native and exotic species, and a rich association of

© The Author(s) 2017
I.D. Rotherham, *Recombinant Ecology - A Hybrid Future?*,
SpringerBriefs in Ecology, DOI 10.1007/978-3-319-49797-6_6

invertebrate fauna. However, to date, there is almost no knowledge of the communities and of its potential when allowed to develop over time. Similarly, many bare ground or early successional communities of urban commons and post-industrial sites develop diverse flower-rich, plant communities with associated invertebrates. As demonstrated by Gilbert (for example 1989), these evolving ecologies reflect both local environmental conditions and human cultural history. However, unless site conditions are extreme (such as waterlogging, acidity, drought, or toxicity for example), such associations are often short-term. It is unknown to what extent these novel ecological communities are replicable between sites, areas, and regions or over time. Many communities converge as tall herb, scrub and secondary woodland take over. There have been almost no rigorous, long-term studies on these dynamic systems.

Similarly, whilst Gilbert (1989, 1992) undertook ground-breaking studies to map recombinant vegetation along urban watercourses, none of the work has been followed-up to examine either associated fauna or long-term dynamics. Nor has there been any detailed survey and assessment of the clearly recombinant and often ecologically diverse aquatic communities now spreading throughout urban and rural areas. There is some monitoring and tracking of higher profile invasive alien species, but little understanding of the emerging communities and ecosystems. Clearly, as noted earlier, some of these associations are novel and appear to be sustained. There is much to do before we gain a fuller understanding of the dynamics of British recombinant ecologies.

The Mechanisms of Change

Ecosystems change over time through the biological processes of successional progressions, of natural species interactions and movement, and by long-term fluctuations in climate and other key environmental factors. Humans then affect these processes through all aspects of land management, especially agriculture and forestry, and through industrialisation and urbanisation. Some effects are wrought directly on the natural resources and others are indirect results of exploitation, such as for example, the release of herbivore populations by the removal of top carnivores. Clearance of vegetation, cultivation of land, drainage and the use of fire, all transform landscapes and ecologies. As human populations increase the effects of settlements and of pollution, remove some species and allow colonisation by others. Finally, people deliberately hunt and exterminate some animals, but move and introduce many other species of animals and of plants. This list is not intended to be comprehensive. However, it is useful to summarise some key impacts of human activities in triggering eco-fusion processes:

Summary of some key processes triggering eco-fusion

1. **Species introduction (deliberate)**
2. **Species introduction (accidental or unofficial)**
3. **Species extinction or displacement (deliberate)**
4. **Species extinction or displacement (accidental or unofficial)**
5. **Pollution**
6. **Habitat destruction**
7. **Habitat creation—novel systems**
8. **Anthropogenic climate change—from micro-climate to more generally**
9. **Species hybridisation (deliberate i.e. in collections or cultivation)**
10. **Species hybridisation (accidental through introductions, mixing, and releases)**

Some British examples of extinctions and introductions:

(1) **Fauna extinction**:
Species removal from the ecosystem: e.g. brown bear, beaver, grey wolf, European lynx, aurochs, polecat (over much of its range but now recovering), otter (over much of its range but now recovered), wild boar, wildcat (England and Wales), pine marten (over most of England and Wales)

(2) **Fauna release**:
Deliberate species introductions: e.g. fallow deer, muntjac deer, edible dormouse, rabbit, brown hare, grey squirrel, numerous fish species for sport

(3) **Fauna release**:
Accidental or unofficial species introductions: e.g. coypu, American mink, brown rat, black rat, feral cat, wild boar, polecat-ferret

(4) **Bird extinctions**:
Species removal from the ecosystem: e.g. golden eagle (England/Wales), white-tailed eagle (reintroduced to Scotland), red kite, common buzzard, peregrine falcon, osprey, hen harrier, raven (all from most of their range but now many have recovered), great bustard, common crane, spoonbill, bittern (over much of its range), plus many other birds of wetlands and similar habitats

(5) **Bird release**:
Huge numbers of game birds over much of the country: common pheasant, red-legged partridge; little owl

(6) **Bird escapees**:
Canada goose, feral pigeon, eagle owl, ring-necked parakeet, grey lag goose, mandarin duck and many others—the biomass of exotic wild species of birds is now a significant proportion of the total

(7) **Reptile release**:
Terrapin, snapping turtle

(8) **Invertebrate release**:
     American signal crayfish
(9) **Flora release**:
     Deliberate species introductions: Rhododendron, Portuguese laurel, Japanese
     knotweed, Himalayan balsam, giant hogweed, numerous water plants,
     sycamore, European larch, Sitka spruce, Scots pine (outside the Scottish
     Highlands), *Sorbus* hybrids, horse chestnut, sweet chestnut and many others
(10) **Flora release**:
     Accidental or unofficial species introductions: e.g. Spanish bluebell, Norway
     Maple, beech (in northern Britain), Montbretia, variegated yellow archangel,
     snowdrop, sweet cicely, and many others

Some examples of species hybridisations:

Fauna:

1. **Red Deer/sika**: Native or feral red deer × alien sika deer where the ranges
   overlap plus extensive backcrossing with the hybrids
2. **Polecat/Ferret**: Polecat × domestic-bred ferret and polecat-ferret—especially
   as polecat re-establishes across its former territory
3. **Cat**: Wildcat × domestic or feral cat—in Scottish Highlands

Flora:

Wild Rhododendron: *Rhododendron ponticum × maximum × catawbiense × macrophyllum*. This is a remarkable example of recombinant ecology whereby the 'wild rhododendron' now spreading over much of Britain, is in fact a complex hybrid swarm of the above species. Furthermore, the original introduction to Britain was from Iberia and an Atlantic-adapted sub-species with limited frost-tolerance. The hybrids were deliberately created in cultivation to give better cold resistance and the resultant hybrids are self-selecting in the wild to favour the frost tolerance genes in colder areas. A further complication is that *Rhododendron ponticum* was native to Britain before the last glaciation and simply never re-colonised.

   Common Montbretia: *Crocosmia aurea × C. pottsii—Crocosmia × corcosmiflora*. This hybrid was deliberately produced in cultivation in France in 1879 from South African parents. It has since spread from garden introductions to become, in some areas, a dominant component of the vegetation. There are varying degrees of back-crossing and a hybrid swarm with potential for adaptation to particular environmental factors.

   Variegated Yellow Archangel: *Lamiastrum galeobdolon* var. *argentatum*. This was almost certainly a Victorian garden plant from a spontaneous mutation to give sterility but hybrid vigorous and stunning variegated leaves. The latter feature that is attractive to gardeners is the reason for its success, along with its ability to grow from stolon fragments. The plant is now thoroughly established in woodlands and spreads into the heart of supposedly semi-natural vegetation.

Willowherb: *Chamerion angustifolium*: In the 1800s in England, this species was small and restricted to a few remote sites on heaths and mountains. Rackham (1986) suggests that the now large and widespread plant resulted from hybrid vigour generated by mixing of North American and British stock around major seaports. The plant is now thoroughly naturalised and dominant in many tall herb communities.

Bluebell: *Hyacinthoides non-scripta* × *Hyacinthoides hispanica*, and various hybrid swarms and back-crosses including with garden 'giant' triploid variants of both parents: Whilst generally recognised in the wild in Britain only at a relatively late date, i.e. the 1980s, a range of highly fertile hybrids is now spreading widely. However, although these plants have the potential to penetrate into the heart of semi-natural, ancient woods, there is little evidence so far that this is happening. Indeed, it may be that the sheer amount of native pollen in such sites means that any exotic pollen does not show up from simple inspection of the phenotypes. This of course does not mean that the genotypes are remaining pure-bred. However, although the Spanish parents have been in Britain and mixing with native since the 1600s, there is little evidence of colonisation deep into native woodland haunts. Nevertheless, observations suggest rapid colonisation of exotic parents and hybrids into enclosure hedgerows aged between 200 and 300 years in areas such as East Anglia. Interestingly, most of these sites lack native bluebell, and this recent spread might relate to a warming climate.

Landscapes Transformed

These historical changes to 'native' ecology were not isolated from other impacts. Indeed, at about this time, the eighteenth and nineteenth centuries, wider landscapes in Britain were traumatised by parliamentary enclosures. Commonland was wrested from commoners, peasants and the poor, and converted into intensive food production units; a process continuing to the present day. Natural or traditional countryside and ecology were swept away by the tide of change (Rotherham 2014a). For example, traditional coppice woods were converted to high forest plantations, and industrialising cities sprawled over the countryside (Rackham 1980, 1986; Rotherham 2014a). In the areas that remained relatively intact, increasingly populated by plants and animals from around the world, lands morphed into leisurely landscapes for the pleasure of landowning industrialists (Rotherham 2014a).

A result of these changes was the transformation from ecology dominated by native 'stress tolerators', to exotic species which were mostly 'ruderal' and 'competitive' plants. This trend was noted for the twentieth century by Davis et al. (2001), but it really began much earlier with landscapes flexing and changing. Disturbance and nutrient enrichment (eutrophication) emerged as dominant influences (Rotherham 2014a).

Transformed Ecologies

An uncomfortable truth based on ecological history and compounding effects of the cessation of traditional and customary countryside practices is a radically transformed ecology. This process increased in significance throughout the 1800s into the late 1900s. The backdrop to this is 'cultural severance' (Rotherham 2009b, 2013b), which equates to the ending of traditional and customary uses, values and management of (mostly) rural and ecological resources. For many landscapes, release from subsistence exploitation of centuries has meant rapid increase in biomass and nutrients, whilst stress tolerant species (often of high conservation value) go into rapid decline (Webb 1986, 1998). The result of modern-day changes is either abandonment or pulses of macro-disturbance. These replace the micro-disturbance associated with traditional management (Rotherham 2009a, 2014a).

Recombinant Ecologies

It is clear that these ecosystem stresses are most obvious in urbanised zones. Here, combined with exotic species as described earlier, these species form new ecological associations, a 'recombinant ecology' (Barker 2000; Rotherham 2014a). The emerging communities are different and distinctive from what went before and long-term trends in vegetation for example, can be recognised at regional scales (e.g. Hodgson 1986). Work by for example Hodgson (1986), and Grime et al. (2007) confirmed the declines and replacements expected from cultural severance and environmental transformation (Fig. 6.1).

In these changed and evolving environments, it is possible to identify long-term indicators of continuity, particularly stress tolerant native species and stress tolerant ruderals. Indeed, there are elements of former, historic landscapes in 'semi-natural' (or eco-cultural) habitats formerly maintained by traditional or customary management, and now, sometimes, by nature conservation. However, even here the cultural drivers of human exploitation over centuries have changed and now often ended (Rotherham 2009b, 2014a). In some cases, there is subtle, long-term blurring of ecology, but in others, changes are rapid and dramatic.

Considering the changes, it is remarkable that so much former landscape and ecology are visible through the modern veneer. Indeed, some elements of the ancient ecosystems are surprisingly resilient, unless swept aside totally by modern mechanisation (Rotherham et al. 2013). However, despite this, there are major issues for conservation such as for example, in Britain, the latter-day recognition of medieval parks as important remnants of the so-called 'Frans Vera primeval landscape' (Rotherham 2007; Vera 2000). For decades, these areas received little recognition or protection. Recent work (e.g. Rotherham 2014a, 2015a, b, c) has suggested how ecosystem components may have persisted from the European

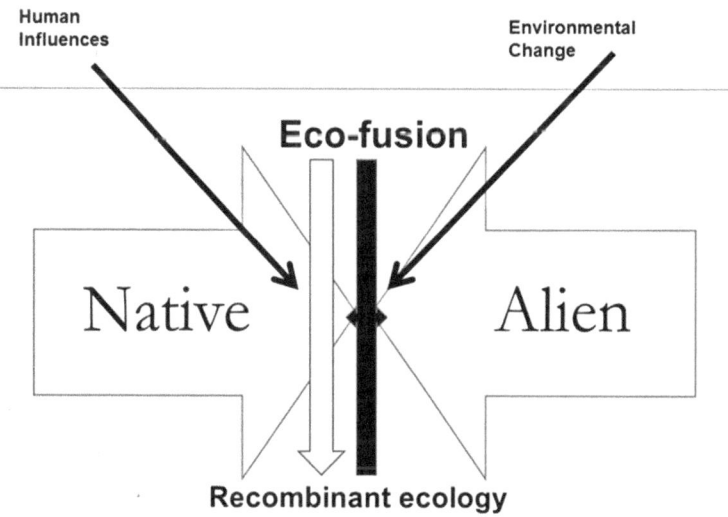

Fig. 6.1 Influences on eco-fusion of natives and aliens to produce recombinant ecologies

primeval landscapes as described in various ways by for example, Vera (2000), Rackham (1986), and Peterken (1996). Furthermore, it is described how these widespread environments and their ecosystems were subsumed into the medieval countryside, with key elements surviving into modern-day conservation reserves. In considering futurescapes and novel ecosystems, these species are especially important constituents of conservation significance. This long-term evolution and transmission through centuries of sometimes-seismic change may be a key to future sustainability. Through understanding the scale and nature of changes, the implications of cultural severance, globalisation, and climate change it is possible to gain some insight into future challenges for humanity and for planetary biodiversity. It is now widely accepted that to have any chance of even slowing the rapid decline of species, we need futurescapes to be 'wilder', bigger and better connected.

The New Wild: Science, Politics, Environmental Democracy and Managing the Land

A recognition of the changes throughout history and of the recombinant natures of past, present and future ecologies, is important in terms of informing debates on wilder landscapes and 'wilding'. One school of approaches is often called 're-wilding' but this is a loaded term since it implies a return for to a formerly wild state, which in practice has not existed for thousands of years. Indeed, when the landscape was 'wild', it was populated by different species in combinations and with environmental conditions that were radically different from those existing today, or that might exist in the future. Nevertheless, finding ways in which to harness recombinant ecologies to futurescapes that secure species conservation and ecosystem services will become increasingly necessary; and this means 'wilder'.

In current debates on wilding, decisions to abandon sites to feral nature or to intervene with planned release of large herbivores, raise many issues. In order to allow trees freedom to regenerate do we fence out and cull wild (feral) herbivores such as the Scottish Highland red deer? These are all human-determined inter-ventions (Rotherham 2014b), and Ayres (2013) for example, welcomed that 'When you let go of control of the land and let nature run its course it is unpredictable, often with surprising and positive outcomes.' This is fine in principle. However, what happens if you get a long-term, dominant stand of bracken rather than the aspirational bluebell woodland so often promised? Should we control feral red deer numbers or let nature take its course of animal starvation, depauperate woods and little tree regeneration? What happens when the last remnants of rich biodiversity are lost? If we are to intervene, then who does it, why do they do it, what do they do, where do they do it, and when do they do it? Furthermore, of course, who decides and who pays?

An understanding of eco-fusion and recombinant ecology helps to inform this debate. Nevertheless, will land managers, conservationists, and even the public, accept exotic plants such as rhododendron, sycamore, larch, spruce, Japanese knotweed, Himalayan balsam and giant hogweed spreading feral, recombinant ecologies across the landscape? Free-willed, recombinant nature mixes these spe-cies with mink, rabbit, grey squirrel, Canada goose, ruddy duck, ring-necked parakeet, signal crayfish, and exotic or feral deer (Rotherham 2014b). Recombinant ecology emerges from ecological fusion processes, but many conservationists remain reticent to accept such changes, and this will most likely be the case with truly feral ecologies dominated by this heady recombinant mix of exotics and native invasives (like birch and bracken).

In this highly charged context, issues of alien invasive, exotic species need to be considered within the wider setting of environmental change, conservation, and politics (Rotherham and Lambert 2011; Rotherham 2016). Whilst largely unrecognised, we already have hybrid ecologies and this situation will inevitably grow, so ideas of recombination and eco-fusion are increasingly significant (Rotherham 2014a).

There is a widespread acceptance that whilst many nature conservation projects are hugely successful and worthwhile, overall, conservation is still failing (Rotherham 2009b, 2014a). Declines in landscape quality and in biodiversity have not been halted, and hence the increasing interest in alternative approaches to managing the environment. The ideas range from major environmental re-construction works, to so-called 're-wilding' (Taylor 2005; Monbiot 2013a) or 'wilding' (Rotherham 2014b). It is further suggested that the response to intractable declines in ecology should be to allow nature to be free and 'free-willed' (Carver et al. 2012; Fisher 2006, 2013). There are aspects of these arguments that have some merit but others are scientifically weak, misinformed (Vidal 2005), and naïve both politically and historically. Most worryingly perhaps is that some of these ideas in reality amount to abandonment i.e. cultural severance (Rotherham 2009b, 2014b), and others lead to over-intensive grazing and damage to sites (Denton 2013).

Simply 'releasing' sites and their ecology from obvious, direct human influence will not achieve the benefits that the proponents suggest (Rotherham 2014a, b, 2015a, b, c). Affected by air pollution and eutrophication, and with communities derived from centuries or millennia of human–nature interactions, much of our landscape and most conservation sites are eco-cultural with bio-cultural heritage (Agnoletti 2006; Agnoletti and Rotherham 2015). In most cases, lacking keystone species like beaver or large carnivores, successional changes will be dynamic and exciting, but not 'natural' in any historical sense. Suggestions that we should abandon nature conservation management and allow 'nature' to be 'free' and 'natural', whilst superficially exciting, are potentially devastating to species conservation. Invasion biology tells us what 'feral' nature may be like, and it is potentially a neo-liberal, market ecology determined by the survival of the fittest in a globalising and eutrophic world. In considering what we mean by 'natural', we need to be clear about the differences between 'process' and 'product' or outcome. Our future will be recombinant and there is no doubt of that. However, this does not mean that we should abandon species to their fates in neo-liberal ecosystems driven by market-force ecology where what Rob Marrs (pers. comm.) has described as the 'ecological thugs' necessarily win out.

Nevertheless, we begin to see some of the remarkable potentials for a freeing up of nature in projects such as Oostvaardersplassen in the Netherlands and say, Knepp Castle estate in southern England. However, these are not 'natural' systems, but a new form of human, culturally determined landscape, and as such part of a toolkit of possibilities. Just like Ennerale in Cumbria, these two landmark projects are actually carefully designed, implemented, and monitored (Rotherham 2013b, 2014b).

Some Concluding Thoughts on Perceptions and Attitudes

History can inform our understanding of present-day ecologies and our visions of future ones. We are now searching for the so-called 'shadow woods' etched in the landscape from perhaps pre-Domesday but still surviving though often unrecognised (Rotherham 2013c). Similarly, many relict heaths and commons hark back to this antique ecology and yet are sadly abandoned and neglected (Rotherham and Bradley 2011). A point to emerge from these observations is that what we value is not necessarily an ecology which is truly native, but one perceived to be so (e.g. Rotherham and Lambert 2011; Rotherham 2014b). Some of our most ancient landscapes still have little protection and often very unsympathetic management. On the other hand, some of the landscape features and their ecology that now passionately protected, such as eighteenth- and nineteenth-century enclosure hedgerows, are actually imposed exotic features. Many of the 'native' oakwoods from which school children carefully collect 'local' acorns to grow and then plant into 'local provenance' woods were actually not native at all. These are frequently, as the estate accounts confirm, imports from Dutch nurseries in the eighteenth century. Indeed, a forester today can often spot the distinctive manifestations of genetic traits that distinguish native trees from Dutch. There are wonderful ancient hedges from pre-Domesday and these cross ancient landscapes to link patches of wood, common and heath, but they are different and distinct from the imposed barriers that separated commoner from common.

It is clear that ideas of ecological stability depend on time-scale but also on presumptions of long-term environmental stasis. These assumed truths are wrong and even considering short periods, environmental conditions and therefore ecologies fluctuate sometimes markedly. Furthermore, human transformation of environmental conditions has changed ecology in the past and continues to do so today but at an increasing scale and pace. Combined with the changed environmental baselines, people also move species of animals and plants around the landscape and across the world. This is constantly creating new potential mixes of ecologies with ecological fusion processes generating novel, recombinant communities. With globalisation (e.g. Hulme 2009) and urbanisation increasing dramatically, these observations have huge significance for nature conservation.

Discussions of new ways to manage nature and the landscape need to engage with the emerging paradigms of environmental and ecological history. Long-term studies of ecological history in Britain, for example, confirm the hybrid nature of our 'native' ecology. They also indicate the increasingly hybrid character of our future nature. This hybridisation, so-called 'eco-fusion', occurs at the level of the community as aliens and natives mix and merge into recombinant ecologies, but also at that of the species as hybrids emerge from both deliberate and accidental fusion.

It is important to set problems of alien invasive and exotic species into broader contexts of environmental change, conservation, and politics (Rotherham and Lambert 2011). Whether or not we like it, we have a hybrid ecology and this will become more so over time. Therefore, concepts of recombination and eco-fusion are increasingly important (Rotherham 2014a). Additionally, understanding inter-relationships between ecology and politics in determining reactions to biological invasions is important. In bringing together science and politics it is necessary to acknowledge that many conservation decisions are not based on '*truths*', often not even on science, and are not objective. Nevertheless, these are subjective decisions based on the best scientific understanding we have, blended with an emotional response to situations based on and twisted by many social, cultural and historical influences (Rotherham and Lambert 2011).

Importantly, this subjective perception applies to both the professional conservation manager and to the wider public alike. Even the language used to define and to describe the issues is loaded with bias, and the decisions are political and social. Why for example do we seek to eradicate Himalayan balsam as a riverside and roadside invader but not sweet cicely an alien from the mountains of central Europe first recorded wild in Britain in 1777? In the Peak District and South Pennines, it is spreading rapidly and its impact is dramatic. So if control is based on science and objectivity, then why one and not the other. *Buddleia davidii* causes millions of pounds of damage to services and buildings, and is now expanding into woods, hedgerows and other habitats such as cliff-tops, but we welcome it as 'The Butterfly Bush'. In contrast, conservationists dislike the wild rhododendron, which they seek to 'bash' and eradicate (Rotherham 2005a, b, 2009a).

It is important that environmental change and its history engage with debates on exotic and native species. In doing this, there are major challenges to current alien species paradigms, and the approaches raise critical issues of perceptions and judgements. To more effectively relate environmental history to past, present and future ecologies needs novel concepts to be developed. Indeed, the application of environmental history research to ecology has triggered novel ideas of recombinant and hybrid ecology, and of eco-fusion. From the emergence of these concepts there arise significant implications for visions of future ecologies, and the work then meshes with ideas of 'wilding' and 're-wilding'. Such ideas are presently debated inspired largely by the ideas of Vera (2000), but now entering wider, more popular audiences through the writing of authors such as Taylor (2005), and particularly Monbiot (2013a, b).

However, concepts of both cultural severance and ecological fusion are important in understanding how past ecologies have changed, and how future ecologies may evolve. At the core of these debates are critical paradigms concerning what is natural, what is cultural, what is native, what is alien, and what is exotic. For the British case study, possible future devolution in the (dis)-United Kingdom raises further issues (see Warren in Rotherham and Lambert 2011) where conservation managers are trying to decide whether a species should be native to England, Wales, Scotland, or some lesser region. In the face of climate change and the inevitable fluxing of species distributions, this is a nonsense and misunderstanding of the serious matters at stake. It is also totally missing the point about the palimpsest nature of historic landscapes and the value of cultural and historical aspects of the environment. Is it relevant that a plant found in Carlisle was not 'native' in Gretna or beyond, and if it spreads north, should it be eradicated?

This is a formalising of the old idea of beech being only native to southern England and so treated as an alien in the northern regions. It has now been found in the early pollen records for North Yorkshire, and so was native anyway. However, a further point is that in the centuries since the closing of the English Channel, beech would colonise northwards regardless, and would now be native in the north as well the south. Another example is the recent discovery of the pollen of Norway spruce in southern England ... from the Bronze Age, (Hugh Milner pers. comm.), and again raising the question of what we consider native or alien in the first place. There is certainly a case for celebrating and conserving where possible, local and regional distinctiveness and character, but regional ecological xenophobia is a dangerous route down which to travel. A denial of environmental change, either natural or human-induced, is unrealistic.

Emerging Paradigms and New Concepts

A major problem in dealing with the apparently simple matters of alien and exotic invaders is in the difficult relationships between conservation and (1) the cultivation of exotic trees for forestry and for amenity, and (2) with farming, horticulture and gardening. In all situations, there is blending of nature and culture that makes many assertions of native or exotic status fraught with problems. Both (1) and (2) are major causes of the undoubted problems caused by alien invasions. Nevertheless, this does not mean that all the impacts are negative, or even that the bad effects are significant or important in all cases and in all situations. There may also be an issue in that many invasions are probably impossible to halt

A lesson of the British experience is also that perceptions of what is a problem, what is alien or native, and even who is responsible for any management or control, vary dramatically over decades and even over centuries. Furthermore, issues of exotic and invasive cannot be separated from the wider fluxes of society, economy and environment; so where and when we examine a particular species, has a huge effect on our perceptions of it as a positive or negative influence on ecology, economy and on society. Since we now argue that our ecological systems are substantially transformed and much of our ecology is a recombinant hybrid, anyway, these matters of perception become ever more complicated.

This is now a truly global phenomenon since historically, as Western imperialism spread globally during the 1800s and into the 1900s, many of these attitudes and a lot of 'native' plants and animals from 'home' were exported. The twin desires to 'improve' and to 'adorn' created many of the invasion and extinction problems witnessed today. However, the human footprint on eco-cultural landscapes and the spread of species around the planet goes back much further. In Europe for example, our recombinant ecologies date back to the very first waves of human migrants across the continent bringing plants and animals with them. As they settled, people brought new cultures, technologies and approaches to land management, and they transformed the landscape. Wherever and whenever human cultures arrived and settled they triggered processes of eco-fusion and generated recombinant ecologies. Indeed, a review of the pertinent literature also suggests that the nature and behaviour of novel ecologies has been addressed in more earnest around the world than in Britain. In the latter, debates on alien and native predominate and the hybrid nature of many ecosystems is unrecognised.

Ecological fusion and recombination have driven ecosystem change and associated biodiversity since life began. Indeed, these processes will continue so long as life is maintained, part 'natural' and part human 'cultural'. To separate humanity from nature is foolish and misconceived since modern-day nature is not natural. In this context, the emerging concepts of eco-fusion and recombinant ecologies are important to understanding environmental history, and history informs our knowledge of ecology. Furthermore, the insight of environmental history is an element missing from many discussions of future landscapes and particularly the potential for so called re-wilding. In the absence of rigorous environmental history and robust science, visions of future landscapes and sustainable, future ecologies remain dangerously misinformed. In this context, the present work provides a conceptual framework in which future landscapes can be considered more effectively is provided by the combined concepts of cultural severance, the eco-cultural nature of landscapes, hybrid ecology, ecological recombination, and eco-fusion. This approach helps identify critical future paradigms and assists in providing guidance for necessary decision-making.

Some of the ideas and approaches to ecosystem novelty developed by authors such as Hobbs et al. (2013) do not transfer easily to the British scene. This is in part because they focus on ideas of pristine wild or historical ecosystems that have been degraded or transformed by relatively recent human activities or influences. As already noted, the ideas of 'wild' or 'natural' in strongly eco-cultural landscapes can be problematic. Indeed, we can argue further that around the world, the human footprint is evidence in most ecosystems, even if not obvious at first site. Through processes of selective extinctions, species introductions, and land management, humans have transformed ecosystems from subtle shifts in dynamics to massive restructuring of ecology. The confusion arises over issues of scales of times and space. The recombinant and hybrid nature of ecology and of ecosystems is relatively easy to apply to the British and European situations, and in this context, an overview framework is useful (Fig. 6.2).

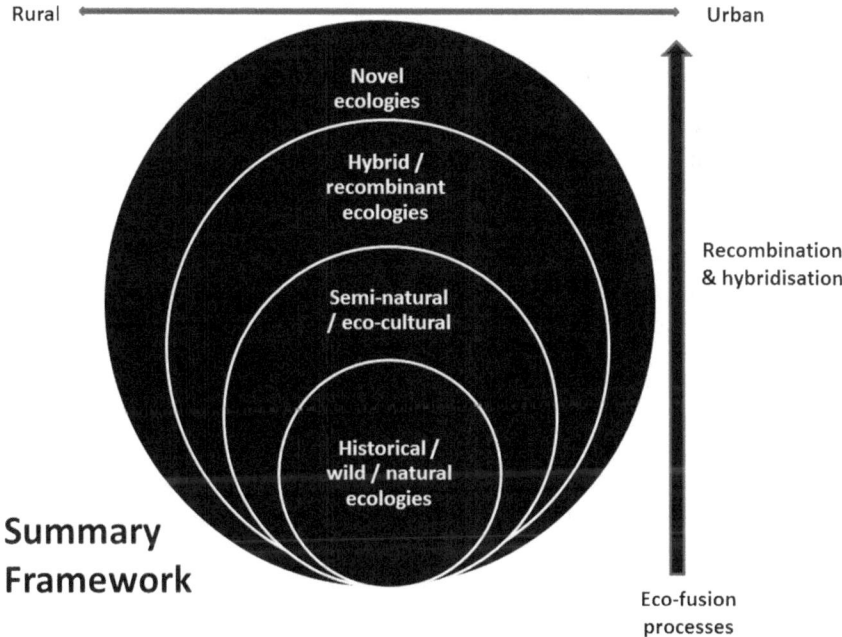

Fig. 6.2 Summary framework

Into this broad framework, we can then place the spectrum of eco-cultural, recombinant, hybrid, and novel ecologies and ecosystems. With a rural-urban spectrum, there are agri-industrial landscapes, agri-forestry, traditional and customary systems, semi-natural/eco-cultural ecosystems, and wild or natural ones. In the urban catchment, there is a range from intensively urban, to post-industrial, to urban commons, to designed and managed spaces, and to encapsulated traditional systems.

Returning to the work of Grime and his colleagues it is reasonable to assume that landscape zones, especially the British uplands for example, will exhibit a greater degree of ecological stasis as available nitrogen and phosphorus restrict many invaders. However, this should be balanced by a recognition of the all-pervading nature of aerial deposition of pollutants like nitrogen, and the impacts of cultural severance as the effects of traditional uses of ecological systems are abandoned (e.g. Rotherham 2008). The ideas are underpinned by long-term studies that demonstrate a separation between upland and lowland Britain in terms of vulnerability to wider changes per se, and to invasibility specifically (e.g. Grime 2005). Furthermore, climate change may complicate matters by shifting growing system dynamics (e.g. Fridley et al. 2016; Stevens et al. 2016). Shifts in soil pH were found to be important drivers in both calcareous and acidic grasslands.

Wilder Futures in a Recombinant World

Almost everything that humanity has done in its relationships with nature, has shifted ecology more and more towards recombinance; species and communities fluxing as people drive the changes. Across the planet now and in Britain especially, the human relationship with nature is changed and changing. With urbanisation, globalisation and climate change continuing, landscapes and ecologies are transformed. For urban centres especially, ecological fusion drives the development of new, recombinant ecologies.

Then, in the future countryside, some areas will continue to be intensively managed as food factories with often entirely constructed and alien ecosystems. Other land may experience a reduction of land-use intensity and often abandonment of traditional land management. Here again, with ecological drivers exerted over centuries ending, it is likely that new, recombinant ecosystems may emerge. However, as established ecology is stressed by factors like climate, eutrophication, and invasive pests and diseases, there will be shifts in ecological dynamics; some species retreating whist others, alien and native, expanding.

One major problem for many native British species in a changing world is that human influences over centuries, but especially since the enclosures period of the 1700s and 1800s, have fragmented the ecological landscape. Habitats have been grossly reduced and what remains is fragmented and often isolated. This is not always a problem for species moving since some can 'hop' between islands or nodes with suitable environmental conditions. However, even these more mobile species need somewhere to 'land' in their search for suitable conditions.

Whilst accepting that future ecologies will be increasingly hybrid fusions of species, but history telling us that to some extent this has always been the case, nature conservation faces major challenges. Indeed, to have any hope of success, we need a bold view of a bigger future landscape. This is not to dismiss or reduce in any way the value of existing nature reserves, since they hold the seed-corn for biodiverse 'futurescapes'. Indeed, this message runs counter to the arguments of many more extreme 're-wilders' who view native species nature reserves as anachronistic. However, these vital reservoirs of biodiversity need to be set in a more joined-up landscape with wilder areas, and these need to be big. Lack of a vision and of a resulting resource that is insufficiently large, not joined up at landscape scale, and not sufficiently wild, will lead to catastrophic loss of species. This process began in earnest in the mid-twentieth century and will continue throughout the twenty-first.

Our understanding of ecological processes is now generally at a stage where we can predict likely changes and directions of ecological evolution. Taking the British landscape as my example it is important to be able to reconcile the multiple demands being placed on an increasingly fragmented and stressed ecological resource. In relation to agricultural policy, Grime (2013), highlights this need with the desire to '… develop a common understanding of how our major ecosystems have assembled and will respond in future to changing patterns of environmental variation and human exploitation'. Invasive and alien species fit into this vision and as Grime and co-workers have noted, the behaviour of such biota can be evaluated with the existing toolkits available to us; non-native, invasive species are not ecologically different from the others. Furthermore, according to Thompson et al. (1995), '…Some of the frequently predicted attributes of exotic invaders (e.g. high dispersal rate, great seed longevity, early reproductive maturity: Roy 1990; Lodge 1993) are not supported…' They go on, '… Invasive species differ significantly from non-invasive species (Thompson 1994), but the attributes of invasive aliens are not unique; most are shared by invasive native species'. In trying to understand the dynamics of ecological recombination and hybridisation, it is important to understand the basic ecological building blocks. With CSR theory, whilst many question aspects of its interpretation, it does help provide a pragmatic approach to understanding species behaviour in our recombinant systems. Grime (2005) presents this in a neat summary in terms of the three broad cornerstones of the CSR triangle:

1. **Monopolists = competitors** such as giant hogweed or stinging nettle—plants which have the capacity to reduce diversity very rapidly by monopolising vegetation.

2. **Wimps = stress tolerators** which are long-lived and slow-growing plants, which tend to occupy very infertile soils.
3. **Ephemerals = ruderals** that are the sorts of plants that are weeds in our gardens, living for a very short time, producing seeds early and dying. They are adapted to heavily disturbed situations.

Whilst CSR theory has been derived mostly from detailed work on flowering plants, its wider application has been discussed and indeed, criticised. Such general use of the approach was advocated quite early in the development of ecological science, by for example, Ramenskii (1938). Recent work (Grime 2013) suggests the approach can be applied to a wide biota and that there may be three-way trade-offs between stress, disturbance, and competition during the lifecycles of particular species. Importantly in considering recombinant ecologies, it appears that CSR functional specialisation of a particular animal or microorganism may be influenced by the CSR equilibrium of a food-plant or host. In terms of invasions into new territories, the greatest functional differences seem to be between invasive and non-invasive species regardless of being native or alien to a locale. This is an important 'take-home message' from Grime's work. It is suggested that the attributes of invasive alien species are strongly habitat dependent and this helps to explain the contradictions in lists of supposed characteristic traits of invaders as presented by authors such as Roy (1990).

British ecological science benefits from a long history of detailed research and theorising, and if we assume invaders share common behaviour traits, whether native or alien, then the science helps both in understanding current trends and in predicting the future. Grubb (1977) discussed the theory of the regeneration niche and this is important in recombinant ecology as it helps explain how a new species might enter the system. Essentially, it is suggested that species that are similar in traits to the mature plant in a community or ecosystem may be able to gain entry through the possession of distinctly juvenile traits that enable them to exploit regeneration opportunities or niches. This is a helpful perspective on recombination.

Ecological and landscape histories can help inform such understandings, but these are hampered by the silo mentality of much research which fails to benefit from genuine inter-disciplinarily. Nevertheless, this future vision is not a matter of '**re**-wilding' since that implies going backwards to a formerly wild state, which in reality never existed. Future visions should look forwards to a new 'wild'. Indeed, as argued here, this will inevitably be a new nature forged of recombinant ecologies and intimate mixes of now native and exotic species delivering ecosystem services and functions. Ecological fusion will generate hybrid ecology much as it has done in the past. Humanity has triggered the drivers of these changes but it also has the potential to create a template on which nature can reconfigure its baseline condition.

References

Agnoletti M (ed) (2006) The conservation of cultural landscapes. CAB International, Wallingford, Oxon, UK

Agnoletti M, Rotherham ID (2015) Landscape and biocultural diversity. Biodivers Conserv 24:3155–3165

Ayres S (2013) The feral book—reintroducing rewilding. ECOS 34(2):41–49

Barker G (ed) (2000) Ecological recombination in urban areas: implications for nature conservation. A Workshop Held at the Centre for Ecology and Hydrology (Monks Wood), 13th July 2000. UK Man and Biosphere Committee, Urban Forum, English Nature, Centre for Ecology and Hydrology, Peterborough, 25 pp. ISBN 185716542X, 9781857165425

Carver SJ et al (2012) A GIS model for mapping spatial patterns and distribution of wild land in Scotland. Landscape Urban Plann 104(3):395–409

Davis MA, Thompson K, Grime JP (2001) Charles S. Elton and the dissociation of invasion ecology from the rest of ecology. Divers Distrib 7:97–102

Denton J (2013) Comment: conservation grazing of heathland—where is the logic? Br Wildl 24(5):339–347

Fisher M (2006) Future natural—the unpredictable course of wild nature. ECOS 27(1):1–4

Fisher M (2013) Wild nature reclaiming man-made landscapes. ECOS 34(2):50–58

Fridley JD, Lynn JS, Grime JP, Askew AP (2016) Longer growing seasons shift grassland vegetation towards more-productive species. Nature (on-line), 16 May 2016, 1–4

Gilbert OL (1989) The ecology of urban habitats. Chapman and Hall, London

Gilbert OL (1992) The ecology of an urban river. Br Wildl 3:129–136

Grime JP (2005) Alien plant invaders; threat or side issue? ECOS 2093(40):33–40

Grime JP (2013) An evo-ecological approach to agricultural policy. Aspects Appl Biol 121:1–10

Grime JP, Hodgson JG, Hunt R (2007) Comparative plant ecology. A functional approach to common British species, 2nd edn. Castlepoint Press, Dalbeattie

Grubb PJ (1977) The maintenance of species richness in plant communities: the importance of the regeneration niche. Biol Rev 52:107–145

Hobbs RJ, Higgs ES, Hall CM (eds) (2013) Novel ecosystems. Intervening in the new ecological world order. Wiley-Blackwell, Chichester

Hodgson JG (1986) Commonness and rarity in plants with special reference to the Sheffield flora. Biol Conserv 36(3):199–252

Hulme PE (2009) Trade, transport and trouble: managing invasive species pathways in an era of globalization. J Appl Ecol 46(1):10–18

Lodge D (1993) Biological invasions: lessons for ecology. Trends Ecol Evol 8:133–137

Monbiot G (2013a) Feral: searching for enchantment on the frontiers of rewilding. Allen Lane, London

Monbiot G (2013b) The Lake District is a wildlife desert. Blame Wordsworth. Guardian, Monday 2 September

Peterken GF (1996) Natural woodland—ecology and conservation in northern temperate regions. Cambridge University Press, Cambridge

Rackham O (1980) Ancient woodland: its history, vegetation and uses in England. Edward Arnold, London

Rackham O (1986) The history of the countryside. Dent, London

Ramenskii LG (1938) Introduction to the geobotanical study of complex vegetation. Selkozgiz, Moscow

Rotherham ID (2005a) Invasive plants—ecology, history and perception. J Pract Ecol Conserv Spec Ser 4:52–62

Rotherham ID (2005b) Alien plants and the human touch. J Pract Ecol Conserv Spec Ser 4:63–76
Rotherham ID (2007) The implications of perceptions and cultural knowledge loss for the management of wooded landscapes: a UK case-study. For Ecol Manag 249:100–115
Rotherham ID (2008) Lessons from the past—a case study of how upland land-use has influenced the environmental resource. Aspects Appl Biol 85:85–91
Rotherham ID (2009a) Exotic and alien species in a changing world. ECOS 30(2):42–49
Rotherham ID (2009b) The importance of cultural severance in landscape ecology research. In: Dupont A, Jacobs H (eds) Landscape ecology research trends. Nova Science Publishers Inc., USA
Rotherham ID, Handley C, Agnoletti M, & Samoljik T (eds) (2013) Trees Beyond the Wood – an exploration of concepts of woods, forests and trees. Wildtrack Publishing, Sheffield
Rotherham ID (ed) (2013a) Cultural severance and the environment: the ending of traditional and customary practice on commons and landscapes managed in common. Springer, Dordrecht
Rotherham ID (ed) (2013b) Trees, forested landscapes and grazing animals: a European perspective on woodlands and grazed treescapes. Earthscan, London
Rotherham ID (2013c) Searching for shadows and ghosts. In: Rotherham ID, Handley C, Agnoletti M, Samoljik T (eds) Trees beyond the wood—an exploration of concepts of woods, forests and trees. Wildtrack Publishing, Sheffield, pp 1–16
Rotherham ID (2014a) Eco-history: an introduction to biodiversity and conservation. The White Horse Press, Cambridge
Rotherham ID (2014b) The call of the wild. Perceptions, history people & ecology in the emerging paradigms of wilding. ECOS 35(1):35–43
Rotherham ID (2015a) Bio-cultural heritage and biodiversity—emerging paradigms in conservation and planning. Biodivers Conserv 24:3405–3429
Rotherham ID (2015b) Times they are a changin'—recombinant ecology as an emerging paradigm. Int Urban Ecol Rev 5:1–19
Rotherham ID (2015c) Relict communities and urban commons—urban distinctiveness, history and sustainable urban diversity. Int Urban Ecol Rev 5:29–38
Rotherham ID (2016) Eco-fusion of alien and native as a new conceptual framework for historical ecology. In: Vaz E, de Melo CJ, Pinto L (eds) Environmental history in the making, vol 1. Springer, Dordrecht, The Netherlands (in press)
Rotherham ID, Bradley J (eds) (2011) Lowland heaths: ecology, history, restoration and management. Wildtrack Publishing, Sheffield
Rotherham ID, Lambert RA (eds) (2011) Invasive and introduced plants and animals: human perceptions, attitudes and approaches to management. Earthscan, London
Roy J (1990) In search of the characteristics of plant invaders. In: di Castri F, Hansen AJ, Debussche M (1990) Biological invasions in Europe and the Mediterranean. Kluwer, Dordrecht, pp 335–352
Stevens CJ, Ceulemans T, Hodgson JG, Jarvis S, Grime JP, Smart SM (2016) Drivers of vegetation change in grasslands of the Sheffeild region, northern England, between 1965 and 2012/13. Appl Veg Sci 19:187–195
Taylor P (2005) Beyond conservation. A wildland strategy. Earthscan, London
Thompson K (1994) Predicting the fate of temperate species in response to human disturbance and global change. In: Boyle TJB, Boyle CEB (eds) NATO advanced research workshop on biodiversity, temperate ecosystems, and global change. Springer, Berlin
Thompson K, Hodgson JG, Rich TCG (1995) Native and alien invasive plants: more of the same? Ecography 18:390–402
Vera F (2000) Grazing ecology and forest history. CABI Publishing, Oxon, UK

Vidal J (2005) Wild herds may stampede across Britain under plan for huge reserves. Guardian:
    Thursday October 27
Webb NR (1986) Heathlands. Collins, London
Webb NR (1998) The traditional management of European heathlands. J Appl Ecol 35:987–990

Photograph: The grey squirrel is a twentieth-century arrival from North America